1924–1927: The Dawning of Quantum Mechanics

Hans-Hennig von Grünberg
Alex Griffiths

1924–1927: The Dawning of Quantum Mechanics

Hans-Hennig von Grünberg
Universität Potsdam
Potsdam, Germany

Alex Griffiths
Berlin, Germany

ISBN 978-3-662-70044-0   ISBN 978-3-662-70045-7 (eBook)
https://doi.org/10.1007/978-3-662-70045-7

Translation from the German language edition: "1924–1927: Der Frühling der Quantenmechanik" by Hans-Hennig von Grünberg and Alex Griffiths, © Springer-Verlag GmbH 2024. Published by Springer, Berlin, Heidelberg. All Rights Reserved.

© The Editor(s) (if applicable) and The Author(s), under exclusive license to Springer-Verlag GmbH, DE, part of Springer Nature 2024

This work is subject to copyright. All rights are solely and exclusively licensed by the Publisher, whether the whole or part of the material is concerned, specifically the rights of translation, reprinting, reuse of illustrations, recitation, broadcasting, reproduction on microfilms or in any other physical way, and transmission or information storage and retrieval, electronic adaptation, computer software, or by similar or dissimilar methodology now known or hereafter developed.

The use of general descriptive names, registered names, trademarks, service marks, etc. in this publication does not imply, even in the absence of a specific statement, that such names are exempt from the relevant protective laws and regulations and therefore free for general use.

The publisher, the authors and the editors are safe to assume that the advice and information in this book are believed to be true and accurate at the date of publication. Neither the publisher nor the authors or the editors give a warranty, expressed or implied, with respect to the material contained herein or for any errors or omissions that may have been made. The publisher remains neutral with regard to jurisdictional claims in published maps and institutional affiliations.

Editorial Contact: Andreas Ruedinger

This Springer imprint is published by the registered company Springer-Verlag GmbH, DE, part of Springer Nature.
The registered company address is: Heidelberger Platz 3, 14197 Berlin, Germany

If disposing of this product, please recycle the paper.

Prologue: The Eight Protagonists of This Story

Imagine a large, bright yellow, sun-drenched garden hall with a view of the adjacent park. It is the summer of 1925, and the eight heroes of our story are gathered together, the two most important sitting on a sofa in the back right corner deep in lively conversation: Werner Heisenberg, 23 years old, and Wolfgang Pauli, 25. Although they are also brilliant mathematicians, in this story they are above all passionate physicists. They are the real drivers, the ones who have new ideas first and are willing to share them. They are fundamentally different in nature and character, like cats and dogs, although they followed remarkably parallel paths in life. Both were introduced to physics by the famous Professor Arnold Sommerfeld at the University of Munich.

On the left at the door to the library are two men, both 22 years old: Pascual Jordan and Paul Dirac. They have both recently graduated, one has already completed his doctorate in Göttingen, and the other will do so in a year's time in Cambridge. They look rather awkward, appear withdrawn, and seem to be lost in thought, but are intently listening in on Pauli and Heisenberg's conversation. They certainly don't speak. Jordan, with his enormous, thick glasses, is self-conscious of his stutter. And the Englishman

Dirac almost never speaks. They are the two mathematicians in this play: enormously talented, they bring into mathematically perfect form what the physicists present to them. Unfortunately, they always come up with identical ideas at the same time. They are almost in competition with each other, and their relationship will be pivotal to the theory of quantum mechanics.

For the time being, that is it for our four twenty-somethings in this story.

Now Messrs. Max Born, 42, and Niels Bohr, 39, enter the room. Both are well-known professors, one in Göttingen, the other in Copenhagen. Max Born heads over to Pascual Jordan, whom he pats on the back in a friendly manner. Born is the head of the Institute of Theoretical Physics in Göttingen, and Pascual Jordan is his assistant. A strong mathematical mind himself, he stands in good company with Jordan and Dirac. The Dane Niels Bohr, on the other hand, goes straight over to Pauli and Heisenberg and, in his loud way, immediately inserts himself into their conversation. A physicist through and through, he is something of a father figure to the two youngsters on the sofa.

Only now do you notice that right in front on the left a young, very elegant Frenchman is sitting with his legs crossed in an armchair, quietly sipping a mocha by himself. It is Prince Louis de Broglie, 32 years old, whom the six gentlemen in the room somehow seem to take little notice of. But now Prof. Erwin Schrödinger, 37, enters the room. He avoids the physicists behind Bohr, doesn't want to go into Max Born's corner either, but goes straight to Louis de Broglie, whom he would like to thank for a brilliant idea.

This, then, is our group of actors. The play lasts from September 1924 to October 1927. We are only in the summer of 1925, but our guests have already memorized their parts for all 3 years and we can ask each to briefly summarize what their role in the story is or will be. Niels Bohr immediately speaks up. Together with Arnold Sommerfeld, he developed an atomic model according to which electrons orbit around atomic nuclei like planets around the sun. But he understood very early on that the model is useless and will have to be replaced by a comprehensive new theory. That is why he recruits younger people for the task and steers them in the right direction. He is like the central coordinator of the entire endeavor. Prince de Broglie reports meekly and succinctly that he contributes the idea that an electron can also be a matter wave. Wolfgang Pauli is third in line; he has the role of the driving force, talks about his principle of exclusion and, above all, the idea of electron spin, which opens doors for the others to walk through. When Werner Heisenberg finally takes the floor, he takes credit for starting the entire story with the decisive impetus for matrix mechanics and later crowned his achievements with the uncertainty principle. With his soft Viennese accent, Erwin Schrödinger briefly but with obvious pride points out that he worked out de Broglie's idea and discovered the Schrödinger equation and with it the wave function. Max Born explains that he, together with Jordan and Heisenberg, sorted out matrix mechanics and came up with the interpretation of Schrödinger's wave function as a probability wave. An awkward pause arises as Jordan is expected to speak next but can't. Finally, Pascual Jordan says very cautiously, suppressing his stutter, that his and Dirac's transformation theory could have combined the approaches of Heisenberg and Schrödinger. And while everyone looks at him with awe, Paul Dirac starts quietly: with his equation that will eventually bear his name, he brought together the theory of relativity and the quantum theory, a feat that allowed explanation of the strange phenomenon of electron spin.

An Englishman, an Austrian, a Dane, a Frenchman, and four Germans gather. Each of the eight guests is given a glass of champagne. An image of the Nobel Prize is engraved on seven of these eight glasses. One will go empty handed. But in the Summer of 1925, only one of the eight knows he is sure to win that award, Niels Bohr. Because he has already received it.

Contents

1924, Introduction: Preliminary Work 1

1925, Thesis: Matrix Mechanics. 31

1926, Antithesis: Wave Mechanics 97

1927, Synthesis: Dualism and Uncertainty 183

**Epilogue: The Eight Protagonists
in the Garden Hall**. 245

Timeline 249

Bibliography. 255

1924, Introduction: Preliminary Work

September 1924, Copenhagen

A train ride from Munich to Copenhagen: what an ordeal! Werner Heisenberg has been on the train for 20 hours and he still has not reached Copenhagen. He should have driven directly from Göttingen instead of making a detour to Munich. But it was worth it. Those beautiful August days in his parents' garden, urgently needed after the stress of the habilitation in Göttingen. Saw some old friends again, wasted some time. What more could one desire? And he will now spend a full 7 months with the world-famous Niels Bohr. And for that, the price of the long train ride from Munich to Copenhagen will be well worth it.

When Wolfgang Pauli and Werner Heisenberg blasphemed, they called Niels Bohr the "Pope of Quantum Physics" and his assistant "His Eminence the Cardinal," referring to his confidant, Hendrik Anthony Kramers. The moody Hendrik Kramers didn't think much of the young

Heisenberg's work and, in his condescending manner, certainly let him know it. The rivalry between Kramers and Heisenberg was obvious to any observer.

Fortuitously Kramers is away all September and the first weeks in Copenhagen will be easy and pleasant for Heisenberg for that reason alone. After all, he will now officially be Niels Bohr's assistant—even if only on a temporary basis for the duration of Kramers' absence. And in Copenhagen that means something.

Ironically, the relationship between Niels Bohr and Werner Heisenberg actually began with the Dutchman Hendrik Kramers. Two years prior Bohr had given his legendary series of lectures on atomic theory in Göttingen, talks that went down in the history of physics as the "Bohr Festival." Everyone, absolutely everyone, who had any standing in physics had come to Göttingen for this Bohr Festival, including the 20-year-old Heisenberg, then still a student at the University of Munich. His sponsor, Professor Arnold Sommerfeld, had paid for his train ticket specifically for this purpose.

It must have been after the third of his seven lectures when Nils Bohr introduced the Stark effect and the latest work by his assistant Hendrik Kramers. Out of nowhere, a complete nobody from Munich spoke up, the student Heisenberg. And in front of the illustrious and astonished audience he tore apart Hendrik Kramer's theory. Unknown to the attendees, Heisenberg had been preparing these remarks for some time, as he was already familiar with this theory from Sommerfeld's seminar. In any case, Heisenberg caught Bohr's eye. In his words:

> *After the discussion ended, Bohr spoke to me and suggested a walk up the Hain mountain in Göttingen. This conversation, which took us all over the wooded heights of the Hainberg, was the first intensive conversation about the basic physical and*

philosophical questions of modern atomic theory that I can remember; and it had a decisive influence on my later life. I understood for the first time that Bohr was much more skeptical of his own theory than some other physicists of the time, e.g. Sommerfeld, and that for him the knowledge of the connections did not come from a mathematical analysis of the underlying assumptions, but from an intensive engagement with the phenomena, which enabled him to intuitively feel the connections rather than derive them. This is how knowledge of nature arises, and only in the second step can it be possible to make what has been known mathematically more precise and make it accessible to full rational analysis.[1]

Mathematics always seems to follow; after you have already understood a problem. You wrestle with the new and the unknown in simple, personal images. Which is exactly what Bohr taught him. After returning from this walk, Bohr is said to have reverently said to his friends about Heisenberg: "He understands everything!" Werner Heisenberg always knew how to impress people quickly. In the fall of 1920, Heisenberg was still just a freshman, but only five years later in 1925 Einstein wrote to his friend Ehrenfest: "Heisenberg has laid a large quantum egg!" At the age of 21, Heisenberg received his doctorate under Arnold Sommerfeld, and at the age of 22 he completed his habilitation under Max Born. By age 23 he was famous.

Following this legendary walk, Bohr and Heisenberg stayed in touch and made plans to work together. Now finally, in mid-September 1924, it was to actually happen: Heisenberg had arrived in Copenhagen for a postdoctoral stay. The subsequent impact of Bohr on Werner Heisenberg's

---

[1] The passages highlighted in italics are always quotations, without this being indicated a second time with inverted commas (Heisenberg W., Schritte über Grenzen, 1971, p. 53).

success in life cannot be overestimated. Many years later, after receiving his Nobel Prize, Heisenberg wrote to Bohr:

> *I know that I learned from you how to do science and that the small contributions that I have been able to make to physics are largely due to the Copenhagen atmosphere in which I grew up and was raised by you. I probably owe most of my current recognition directly or indirectly to you. So, I thank you very much for everything you have done for me.*[2]

October 1924, Cambridge

The *London, Edinburgh and Dublin Philosophical Magazine* and *Journal of Science* published an article[3] by the 26-year-old Edmund Clifton Stoner titled "The distribution of electrons among atomic levels." Stoner works at the famous Cavendish Laboratory in Ernst Rutherford's group. He is not a particularly successful experimental physicist, but he has developed a passion for the work of Niels Bohr and the emerging quantum theory. In this work, he challenges the great Bohr and maintains that Bohr's distribution scheme of electrons on filled subshells is based on somewhat arbitrary arguments. He, Stoner, has a better idea about the rationale for the structure of the periodic table. It is only a small observation. And if the famous Arnold Sommerfeld hadn't found out about it and spread the news, it would probably have been simply lost.

---

[2] Letter to Niels Bohr dated November 27, 1933.
[3] Stoner (1924).

November 1924, Hamburg

Wolfgang Pauli initially let the year 1924 pass without dealing any further with quantum theory, which at this point was still in its infancy. He found his previous work on the anomalous Zeeman effect extremely unsatisfactory and needed to gain some distance. But when he was asked to write a longer article on quantum theory in the fall of 1924, he quickly returned to this problem: the anomalous Zeeman effect to which he had already sacrificed so many years of his life.

What is anomalous about this effect? Atoms in a magnetic field show a splitting of their spectral lines, which depends directly on the strength of the magnetic field. A big field results in a big split, a small field results in a smaller one. Even for a classically trained physicist this is no big surprise. The electrons have an orbital angular momentum, and the angular momentum of a charged particle is always accompanied by a magnetic momentum. The orbiting electrons behave like small bar magnets in an external field and align with it. Therefore, they have different energies which ultimately explain the splitting of the spectral lines; the distances between them have something to do with these energies. These splits can even be understood in the fairly simple Bohr-Sommerfeld model—the state-of-the-art model of quantum theory at that time. The Bohr-Sommerfeld model already has two quantum numbers and a third quantum number can simply be added to the description, namely the magnetic quantum number of the orbital angular momentum. If all observations fit well into the Bohr-Sommerfeld model, the effect is referred to as the normal Zeeman effect. But unfortunately—and here's the catch—this only works in very few cases. Most of the time the lines split into many,

many more lines than one would expect in the classical view. We then call this effect the anomalous Zeeman effect.

Pauli struggled with this for many years. As he wandered aimlessly through the streets of Copenhagen, where he spent his postdoctoral period with Niels Bohr, a colleague spoke to him: "You look very unhappy"—"How are you supposed to look happy when you're thinking about the anomalous Zeeman Effect," he answered. Funny and yet also meant seriously, because it really was making him miserable!

The anomalous Zeeman effect cannot be understood without the concept of an electron spin. An electron doesn't only have a magnetic momentum because of its orbital angular momentum, but it also has its *own* magnetic momentum, intrinsic to itself, so to speak. And since the physicist can never think of a magnetic momentum without an angular momentum causing it, the electron not only has its own magnetic momentum, but also its own angular momentum, which is what we call "spin." In essence, the term spin refers to the intrinsic angular momentum (and thus the intrinsic magnetic momentum) of each electron. But there are many other electrons in an atom and these also have their own magnetic momentums due to their spin and the orbital angular momentum. If you want to understand how an atom with many electrons behaves in a magnetic field, then you essentially have to understand how all these partial magnetic momentums are put together to form the total momentum, a truly mammoth undertaking!

All of this means that the anomalous Zeeman effect was exactly that experimental observation that drove classical physicists to their wits' end! So, if Wolfgang Pauli was walking miserably through Copenhagen, then his misery was that of the entire physics profession at the time, which had to admit that a whole world of terms, models, concepts, and theories awaited discovery and that something essential had so far eluded them, of which spin was the prime example.

But it is precisely this spin that Wolfgang Pauli is now slowly tackling. The crucial clue comes from his teacher Arnold Sommerfeld, namely a reference to the article by Edmund Clifton Stoner. Something suddenly dawns on him and on November 10, 1924, he writes to the experimental physicist Landé in Tübingen about "a strange idea concerning the Zeeman effect." What he means by this becomes clear in his second letter to Landé, dated November 24th:

> *In the case of alkalis, the luminescent electron alone creates complex structures such as the anomalous Zeeman effect. There is no question of any involvement of the noble gas atom residue (even with the other elements). The luminous electron manages in a mysterious, non-mechanical way to run in two states (with the same k) with different impulses.*[4]

The alkali metals are the elements lithium, sodium, potassium, rubidium, cesium, and francium. They all have the same structure, namely a full electron shell as the atomic body—like that of a noble gas atom—plus a single electron on top governing the atom's entire chemistry, namely the so-called luminescent electron. Here in this letter to Landé dated November 24th, Pauli recognizes that the anomaly of the Zeeman effect in the alkali metals arises entirely from the single luminescent electron and that it exists in a "non-mechanical way" in two states, what we nowadays call spin-up and spin-down. So, this is the early dawn of the concept of spin: Pauli understands that an electron exists in two states, that this is the basis of the anomalous Zeeman effect

---

[4] Klein and Toomer (1979, p. 177).

and that ultimately the number of states in which an electron can be in an atom is doubled.

November 1924, Göttingen

His first article. Finally finished! And tomorrow he will send it to the *Zeitschrift für Physik*.[5] Pascual Jordan strokes the cover of his manuscript tenderly. Max Born had suggested the title to him, but he wasn't sure: "On the theory of quantum radiation."[6] That sounded like a deeper, more fundamental insight than the one he had had. And in order not to seem like an impostor, he counterbalanced this by starting the article with the modest words: "An attempt is made to show that…", which Max Born found terrible. He deemed the word "attempt" silly, almost a little submissive. Jordan didn't just "try," he succeeded. With great results! Born repeated this several times in the doctoral examination, rated his doctoral thesis highly, and then suggested publication. As his boss, he was very happy with him.

And in Max Born, Jordan had found the right mentor. He liked him. So the move from the Hannover University of Technology to the large and venerable University of Göttingen 2 years prior had been worth it after all. But his father had been vehemently against this change. If Ernst Pascual Jordan had had his way, his son would have stayed in Hannover and studied architecture. The father was a man of strong opinions about culture and society, a respected painter, a leading member of the Hanover Artists' Association, and himself a professor at the Hannover University of Technology. Ernst Pascual Jordan was

---

[5] Founded in 1920, the Zeitschrift für Physik was the most prestigious physics journal in the world at this time.

[6] Jordan P., Zur Theorie der Quantenstrahlung (1924).

conservative through and through and had nothing but contempt for the new, avant-garde movements like Modernism and Art Deco. How strangely experimental, how unrepresentative this new modernity was, which was so popular with the city's youth. His students, on the other hand, complained about the stuffy old professor who only taught "boring and conservative landscapes."[7] The younger Pascual thought they were idiots. His father was a fantastic painter; why couldn't they admit that? His work was so realistic that once a pigeon flew through the window of his studio and tried to sit on a crucifix that he had just painted. They should learn to do that first before complaining about his father's conservatism!

Pascual Jordan wanted to please his father, but studying architecture... no, he just didn't want to. His mother, Eveline, intervened, and insisted that the boy be allowed to do what he wanted. And if he wanted to study mathematics or physics, that was completely fine. And Eveline knew what she was talking about. She was a gifted mathematician herself and had helped the young Jordan be a top performer at school, getting him books on differential and integral calculus, as well as books by Ernst Mach. She had even learned Latin herself just so that she could help Pascual with his schoolwork. But above all, she was concerned about his spiritual well-being. Through his mother's influence, Pascual Jordan was deeply religious throughout his life and a devout Protestant.

And here in his doctoral thesis, which tomorrow he will send to the *Zeitschrift für Physik*, all of these formative influences have come together and combined in the most beautiful way to form a new whole. Here, in clear mathematical form, one can recognize something of his father's strict artistic conservatism and his impressive perspective skills. The

---

[7] Dahn (2019, p. 45).

work also has something of the careful, ambitious, and hard-working mother who has passed on to him her self-discipline and mental sharpness. And of course, the cosmopolitanism of Max Born and the mathematical precision of Richard Courant are reflected, the man he had helped to write the book "Methods of Mathematical Physics" the year before. With this work, Pascual Jordan is sure that, at the age of 22, one phase of his life is coming to a close. He breathes a sigh of relief. Now the next phase can begin.

November 1924, Paris

Finally, in November 1924, the Physics Faculty of the University of Paris met to hear the defense of Louis de Broglie's doctoral thesis, a full year after the thesis was formally submitted. A work with the beautiful name: "Recherches sur la théorie des quanta," or "Investigations into Quantum Theory." Why did the review take so long? The committee that had to decide on the submitted doctorate consisted of de Broglie's actual doctoral supervisor, namely Paul Langevin, the mathematician Cartan, the physicist Perrin, and the mineralogist Mauguin. Although all four praised the originality of de Broglie's hypothesis, they were more than skeptical as to whether the effect postulated in the work could ever really be measured. Langevin then sent the work to his friend Albert Einstein and asked for his opinion. And his answer was a long time coming. After months, Langevin follows up and finally gets a reaction from Einstein, who writes enthusiastically: "De Broglie has lifted a corner of the great veil."[8] Einstein was one of the few who immediately understood how fundamental de Broglie's idea was. Five years later, the rest of the world

---

[8] Kubli (1970, p. 28).

understood it and de Broglie received the Nobel Prize for his doctoral thesis.

Asking Einstein about de Broglie's idea was actually relatively obvious. Because the ideas of Einstein and de Broglie relate to each other like yin and yang. What Einstein is to light, de Broglie is to the electron. In 1905, Einstein developed the concept of a photon and attributed particle character to light. Depending on the type and implementation of an experiment, the light sometimes appears as a collection of point-like particles and sometimes as an electromagnetic wave field. This is the so-called "wave-particle dualism." This puts a certain amount of stress on the human imagination: a wave is a wave. A particle a particle. How can something be both?

In his doctoral thesis, De Broglie reversed Einstein's train of thought. Einstein draws conclusions from the wave to the particle. De Broglie from the particle to the wave. Einstein deduces the photon from the frequency of the light wave. De Broglie deduces the existence of a wavelength from the momentum of a particle, the electron. Einstein considers light, which has no mass. De Broglie looks at electrons, which do have mass. Not only do we have to allow particle character and wave character at the same time for light, but also for particles with mass. An electron, which in 1924 was seen by the entire world of physics as a particle and only as a particle, is now given a wavelength and a frequency by de Broglie because the electron as a particle can also be a wave at the same time with its typical wave effects such as interference and diffraction. 1924 is, therefore, the birth year of the concept of what is called the "matter wave." The fact that you have to treat matter particles exactly like light quanta is what is actually new about Louis de Broglie's theory, the great achievement with which he generalized Einstein's light quantum theory in a decisive way. From here a direct path leads to the later discoveries of Erwin Schrödinger, who will enter our story only at the end of 1925. This was an incredibly brave step by de Broglie. The venerable physicist and mathematician Lorentz is claimed to have said in horror in a conversation with Planck:

> *These young people take it too lightly to set aside old physical concepts.*[9]

Yes, the time had come for bold steps. In the same month, the universe expanded in a completely different way. In 1923, Edwin Hubble was able to prove that the Andromeda Nebula lies far outside our Milky Way Galaxy and therefore forms its own galaxy. Hubble had presented his results to the American Astronomical Society in November. What was thought since ancient times to be a nebula had spurred many competing explanations, including that of Immanuel Kant, who correctly thought that the nebulous stars were star systems similar to the Milky Way. Hubble's discovery was simply astonishing. Our Milky Way is just one of many galaxies! Our universe is in fact much, much larger than the Milky Way system! What an immense expansion of our horizons! A radical remeasurement of the boundaries of our universe. Is it a coincidence that scientists began to explore both the large and the small limits of the universe, that they tried to understand the laws of the atom at the very moment they became aware of the expansion of the universe?

As boldly as Edwin Hubble advanced at one end of the scale, Louis de Broglie simultaneously advanced at the other end: as a nobody, presenting such a novel idea to the world in a doctoral thesis! How audacious to want to go further

---

[9] Planck (2001, p. 73).

than Einstein, who could have postulated it that way himself if only he had seen it.

December 1924, Berlin

Einstein dwelled on de Broglie's work for a long time. He wrote to Lorentz, who was horrified by it, in December 1924 explaining what so fascinated him about Louis de Broglie's matter wave:

> *I believe this is the first faint ray of light on this worst of our physical mysteries. I found a few things that speak for its construction.*[10]

But who was this de Broglie? The de Broglies were a very well-known family in France. In their ancestral gallery, the title "Marshal of France" can be read three times, the job title "high-ranking diplomats" is listed three times. There was a general and a bishop, and two de Broglies were ministers and even prime ministers of the French Republic. And now we add two famous physicists: the brothers Maurice and Louis de Broglie.

While Louis achieved world fame as a theoretical physicist, Maurice was a gifted and avowed experimental physicist who worked on X-ray spectroscopy. Maurice was by far the most important person for Louis throughout his life. His influence on Louis' development cannot be overestimated. As a 29-year-old officer Maurice left the navy, against the stubborn resistance of his family, to study physics in Paris with the famous Paul Langevin. He received his doctorate under Langevin's supervision and remained

---

[10] Letter from Albert Einstein to H.A. Lorentz dated December 16, 1924.

associated with him throughout his life, even later inheriting his academic chair.

The pair of friends, Paul Langevin and Maurice de Broglie, also made contributions to a very famous series of specialist conferences: the so-called Solvay Conferences. But what are these Solvay Conferences? In 1910, Ernest Solvay, a major Belgian industrialist, met Walther Nernst, then a professor of physical chemistry at the University of Berlin (later known as the Humboldt University). Nernst succeeded in getting Solvay to sponsor a whole series of international physics conferences to be held in Brussels. It was important to both of them from the outset that only the most fundamental questions should be addressed and that only the most famous personalities should attend these conferences: summits in physics with a maximum of 25 physicists and chemists. Still held today, these conferences are the G20 of science. The first conference title came from Nernst himself: "Theory of Radiation and Quanta," chaired by Hendrik Antoon Lorentz. Maurice de Broglie was commissioned by Langevin to write down all the lectures and results in detail. These transcripts were then published, which helped this first conference to become an extraordinary success.

In November 1911, Maurice de Broglie returned from the first Solvay Conference. Physics books are still raving about this conference today, because such an incredibly exciting new beginning in physics never again emerged as clearly and as purely as it did at these exclusive conferences. Maurice understood that the discovery of quanta opened the door to a huge, new, and still completely unexplored wonderland. He must have enthusiastically raved to his little brother Louis de Broglie about it, giving him his notes from the conference. "There, brother, read this. Something big is happening! A whole new physics is emerging! You are young and bright and just starting out. Study physics and get involved with the quanta!" And it was through these 400 pages of notes that Louis de Broglie first came into contact with the emerging field of quantum physics, deciding then and there to devote himself entirely to quantum physics. He breaks off his engagement, banishes the history books he had previously studied from his shelf, reduces his social contacts to a minimum, and from then on devotes himself solely to mathematics and physics.

December 1924, Hamburg

On December 2nd, the editorial team of the *Journal of Physics* received an article[11] by Wolfgang Pauli. The title: "On the influence of the velocity dependence of the electron mass on the Zeeman effect." It is one of those publications that one rarely finds today, as it reports a negative result: something that could have been but is not. What could have been? If one assumes the Bohr-Sommerfeld atomic model is correct and electrons indeed orbit around nuclei, then electrons do so at a significant speed. But according to Einstein's theory of relativity, masses change at very high speeds and Wolfgang Pauli calculates this effect for the luminescent electrons of alkali metals in this paper. In fact, he thinks, this should have a significant effect on the anomalous Zeeman effect and furthermore a much larger effect on alkalis with higher atomic numbers, such as cesium, than on alkalis with small atomic numbers such as lithium. But it is precisely this postulated phenomenon that Wolfgang Pauli's colleagues, the experimental physicists Alfred Landé and Ernst Back, were unable to confirm. And so Pauli now

---

[11] Pauli W., Über den Einfluß der Geschwindigkeitsabhängigkeit der Elektronenmasse auf den Zeemaneffekt (1925).

knows that the anomaly in the anomalous Zeeman effect is not an effect of the theory of relativity and is not due to the increase in mass due to higher speeds. Furthermore, he now knows for sure: the Bohr-Sommerfeld model of electrons orbiting nuclei no longer holds. It is an outdated model. But how can the Zeeman effect otherwise be explained? The still speculative answer to this question is what makes the publication so important:

> *In particular, for the alkalis, the momentum values of the atom and its energy changes in an external magnetic field are essentially considered to be a sole effect of the luminous electron, which is also considered to be the seat of the magnetomechanical anomaly.*[12]

And he continues:

> *According to this point of view,* [the anomalous Zeeman effect] *comes about through a peculiar, classically indescribable kind of ambiguity in the quantum theoretical properties of the luminous electron.*

And with this word "ambiguity" the concept that will ultimately lead to the notion of electron spin has arrived in the world. In essence, spin was born half a year before quantum mechanics. Unequivocally Pauli states that the luminescent electron can be in two states with different angular momentum. If he had added a clear interpretation and, for example, written about the self-rotation of the electron, sometimes in one direction, sometimes in the other, if he had added just one such paragraph, he would have gone down in history as the discoverer of spin. An abstract approach to ideas invariably prevents other people from catching on and

---

[12] Ibid.

getting on board. A year later two students from the Netherlands made up for his clear, but unfortunately incorrect, interpretation: Samuel Goudsmit and George Uhlenbeck are now considered to be the discoverers of spin.

Wolfgang Pauli refrained from any descriptive interpretation of his results, not only in this paper but in principle. He had internally accepted that the model of an electron flying around nuclei in fixed, predictable orbits simply did not correspond to reality. And he didn't just want to do away with this descriptive model, he wanted to say goodbye to every descriptive model in the physics of the atom. How foolish that we set off into the world of atoms with these childish ideas in our heads, only to have to realize again and again that these very models are actually just getting in our way. Why should a model derived from the views of the macroscopic world be able to somehow describe the conditions in the microscopic world? Isn't that a completely different world? With completely different laws? He wrote to Arnold Sommerfeld about this on December 06, 1924:

> *The modelling concepts are now in a serious crisis of principle, which I believe will ultimately end with a further radical intensification of the contrast between classical and quantum theory. We now have the strong impression with all models that we are speaking a language that is not sufficiently adequate for the simplicity and beauty of the quantum world.*[13]

And he wrote to Bohr a week later:

> *However, we must not try to force the atoms into the shackles of our prejudices (which, in my opinion, also include the assumption of the existence of electron orbits in the sense of ordinary*

---

[13] Letter to Arnold Sommerfeld dated December 6, 1924, see Klein & Toomer (1979, p. 182).

kinematics), but rather we must, conversely, adapt our concepts to experience.[14]

If you know how close Werner Heisenberg and Wolfgang Pauli were in contact with each other, then this sentence, in particular, reads like an instruction to Werner Heisenberg, who, just 6 months later, created quantum mechanics in Heligoland, starting from just this assumption: electrons in the atom do not move in fixed orbits around the nucleus. We have to completely abandon the idea of electron orbits as no one will ever be able to observe them. And if you can't observe them, they don't exist. So you can see that Wolfgang Pauli gives the clever Werner Heisenberg these two stage directions:

1. Please, no models which are based on vivid ideas just because it is pleasant to be able to imagine the processes.
2. And especially no more electron orbits!

And, of course, that was the script with which Heisenberg became successful. Equally clear: when he discovered the spin, Wolfgang Pauli simply did not want to get involved in a beautiful but always somewhat doubtful interpretation of what he instead called just the "ambiguity" of electrons.

Thus, this was the situation at the end of 1924: Pauli sets out to discover the spin and gives his friend Heisenberg the important stage directions that will help him to create the actual quantum mechanics. And in order to follow the story over the next few years, you have to imagine a braid that is in typical fashion braided with three strands. The first and earliest strand is Wolfgang Pauli and the discovery of spin. The second strand is Werner Heisenberg and his Heligoland

---

[14] Letter from Wolfgang Pauli to Niels Bohr dated December 12, 1924, see Klein and Toomer (1979, p. 186).

discovery that electrons are not to be understood in terms of orbits but in terms of matrices. And the third strand will only come into play in 1926 when Erwin Schrödinger develops his wave mechanics. First, Schrödinger's strand is connected to Heisenberg's but only much later, in 1928, Paul Dirac manages to braid the third strand, the one with spin, in order to truly complete the braid.

Incidentally, it is not clear whether Heisenberg and Pauli were really friends in the true sense. Heisenberg, born in 1901, and Pauli, born in 1900, are the two boy wonders of this story. They have the same three academic mentors, world-famous theorists who trained them with absolute dedication and attentiveness. Professors Arnold Sommerfeld from Munich, Max Born from Göttingen and Niels Bohr from Copenhagen, together transformed the two talented students into absolutely exceptional physicists.

I learned optimism from Sommerfeld, mathematics from Göttingen, physics from Bohr.[15]

Heisenberg wrote in later life. In the very first weeks of his physics studies in Munich, when Heisenberg was still just 18 years old, Arnold Sommerfeld introduced him to his assistant Wolfgang Pauli, then 20 years old. Heisenberg and Pauli only spent two semesters together in Munich before Pauli moved to Born in Göttingen, followed later by Heisenberg. Criticizing each other with remarkable vehemence, they would write letters to each other for the rest of their lives. Heisenberg on Pauli:

---

[15] Hermann, Heisenberg (1976, p. 28).

Throughout his later years, as long as he lived, he played the role of an always welcome, if very harsh, critic and friend for me and for what I was trying to do scientifically.[16]

Pauli was in pretty much every respect the exact opposite of Heisenberg. While Heisenberg studied in Munich in a well-mannered, diligent, and serious manner, ideally outside in nature, Pauli enjoyed the Munich nightlife, hung out in cafes and bars, went to bed late, slept in and actually only came to lectures when they were already over to quickly write down what Sommerfeld had said with a glance at the board. "Good morning, you apostle of nature!" Pauli used to greet Heisenberg during their days together in Munich. "Good afternoon," Heisenberg replied. But the contrasts went far beyond that: one of them, Pauli, was full of scruples and therefore rather hesitant, the other, Heisenberg, naive and also a little hesitant but invariably a doer. What discoveries Pauli could have made if only he had been less cautious. But in the end, it was probably this contrast that made the two physicists defer to each other again and again. Heisenberg, the student, and Pauli, the assistant—the two were a team, from their very first meeting until Pauli's death in 1958, a black horse and a white horse, hitched together to the same wagon. This carriage was the emerging quantum mechanics, pulled forward in the 1920s by the two of them with great passion and involvement. A unique stroke of luck for physics.

While both had big egos, Wolfgang Pauli's self-confidence bordered on insolence and arrogance. After Einstein finished a talk in a packed lecture hall, Paui, still a student at the time, remarked loudly: "What Einstein is saying isn't all that bad." He was a child prodigy, and he knew it. Not only did he feel superior to everyone, he *was* superior. And he

---

[16] Heisenberg W., Der Teil und das Ganze (2022, pp. 35–36).

stood up to the great eminences early on. There are plenty more stories about Pauli. Indeed, so many that Valentin Telegdi, a professor at ETH Zurich, could organize an entire evening event in 1983 of Pauli anecdotes. The event has been handed down in writing and one story, in particular, beautifully demonstrates his cheekiness:

Pauli, still a student at the time, and Ehrenfest, already a respected physicist, had both contributed articles to the Encyclopedia of Mathematical Sciences—Pauli the article on relativity theory, which is still worth reading today, and Ehrenfest (together with his wife) the article on statistical mechanics. While Ehrenfest gave a lecture at the German Physical Society, Pauli frequently interrupted him with critical comments. After the lecture, Ehrenfest approached Pauli and said: "Well, Mr. Pauli, I have to say, I like your article better than I like you!" to which Pauli replied: "For me, Professor, it's just the other way around…"[17]

By the way, this article in the "Encyclopedia of Mathematical Sciences" catapulted Pauli instantly to fame. The background: He was barely 21 years old and only in his third semester as a physics student at Munich, working with the same famous Arnold Sommerfeld who helped many other famous physicists take their first steps. And just as he had supported Heisenberg, Sommerfeld knew he could do something with the extraordinary talent of Pauli, whom he had discovered very early on. This was probably Sommerfeld's greatest ability in the first place: discovering people! In order to properly challenge such a brilliant mind, Sommerfeld entrusts his student Pauli with the task of writing a summary article about Einstein's theory of relativity for the "Encyclopedia of Mathematical Sciences." And this isn't just any old journal, but a highly respected encyclopedia.

---

[17] Enz & von Meyenn (1988, p. 111).

What a courageous decision by Sommerfeld to entrust such an honorable task to a third-semester student! But bold steps usually pay off. Pauli busies himself with this task for almost a year, filling 237 pages with 394 footnotes and capturing the theory of relativity in such an exemplary and clear manner that his summary is still recommended reading for students today. Albert Einstein, in particular, who did not want to write the summary himself, was very impressed:

> *Anyone who studies this mature and large-scale work would not believe that the author is a man of twenty-one. One does not know what one should admire most: the psychological understanding of the development of ideas, the certainty of mathematical deduction, the deep physical view, the ability to provide a clear, systematic presentation, the knowledge of literature, the factual completeness, the certainty of criticism.*[18]

So that was Wolfgang Pauli.

New Year's Eve 1924, Somewhere in Germany

New Year's Eve, the last day of the year, and so 1924 comes to an end. One draws the year to a close by reading the newspaper. The festivities can wait. The party doesn't start until 7 pm, so getting ready around 6 leaves plenty of time to see what has happened over the last few months.

In the *Niederrheinisches Tageblatt*, the following small note can be found from September 18, 1924:

---

[18] Einstein A. (1922).

Waldniel, 15 Sept. Yesterday, Sunday afternoon, Miss Teusch, a member of the Reichstag from Cologne, spoke to a large Centre meeting about the world-historical significance of the Reichstag session of 29 Aug. 1924 concerning the vote on the Dawes Plan and the parties' position on it. Miss Teusch convincingly explained the guidelines of the Centrist policy as a policy of Christian ideology and a united Germany. The rapturous applause after the brilliant, two-hour speech proved to the speaker what an impression she had made on the assembly with her remarks.[19]

The Weimar government struggled to keep up with reparations payments owed to France, an ongoing challenge since the Treaty of Versailles. In response, the French occupied the Ruhr area—and triggered a wave of passive resistance in Germany. And so the USA brokered the Dawes Plan, whereby they would lend the Germans $200 million. This was intended to stimulate the German economy and make reparation payments to France. The Dawes Plan was signed in mid-August.

We moved on from Germany by way of Saxony to India, which is in the sights of the *Sächsische Volkszeitung* on September 28th:

It is with the deepest sense of shame that one reads the words of the Indian Mahatma Gandhi: "modern civilization is the black age, the age of darkness. It makes the goods of the body the only aim of life. It does not concern itself with the goods of the spirit, the soul. It makes the Europeans foolish. It makes them beholden to money, incapable of peace and of any inner life. It is hell for the weak. It undermines the viability of peoples. The civilization of the West must be driven out!" Shouldn't all friends of the people stand together to erase this disgrace?

---

[19] These and the following quotes are taken from the specified newspapers in the German newspaper portal and can be found at https://www.deutsche-digitale-bibliothek.de/

In February 1924, Mohandas "Mahatma" Gandhi—imprisoned by the British for sedition—was released after 2 years in prison. He is suffering from appendicitis and total exhaustion from his many hunger strikes. Back from South Africa since 1915 and inspired by his experiences with Apartheid, he propagates a strategy of non-violent resistance against colonial rule, takes part in numerous hunger strikes and marches and calls for a boycott of British goods. The success of this strategy lands Gandhi in prison. And now, after his release, the movement behind Gandhi really gathers steam. Complete independence is now the goal, making Gandhi world famous in the process.

Moving to the west of the Reich, to Bonn, where the cover of the *Generalanzeiger* of October 15th features the bright headline: "The victory of our aviation technology!" Finally, a victory for the battered Germans, albeit still a very dubious one. The transfer of the Zeppelin LZ 126 to America is a reparation payment from the Germans to the USA.

> *Close to its destination. By the time these lines leave the printing press, the Zeppelin will be only a few nautical miles from Lakehurst. From the numerous radio messages and cable dispatches, which show the individual phases of the giant airship's great ocean voyage, it can be seen that the airship had to fight hard against the elements. While the original course was more or less maintained until after the Azores, it had to be altered from then on as it moved towards the Bermuda Islands and then the course was soon chosen to the south and south-west and finally more to the north towards Sable Island. If the last reported course does not have to be changed again due to weather conditions, LZ 126 will probably take a sharp westerly course directly towards Lakehurst after leaving the dangerous air pressure area at increased speed in the final phase of its journey.*

The Zeppelin, also known as the "reparations airship," was built in Friedrichshafen over the last 2 years and would later become the most successful airship in the service of the US Navy. On October 15th, after an 81-hour and 8000-kilometer journey, it arrived safely in Lakehurst, making it the second airship to cross the Atlantic. People in the Zeppelin company must have been quite relieved because the company had to put up all of its assets to cover liability—the risk was so high that no insurance company was willing to cover it.

"Dissolution of the Reichstag" reads the front page of the "*Deutsche Allgemeine Zeitung*: Greater Berlin edition, Morgenblatt"on October 21, 1924. The article explains that:

The following is officially announced: "The efforts of the Reich Chancellor to expand the current Reich government in order to give it a secure majority in the Reichstag to continue its current policy have finally failed. As a result, the Reich Chancellor, in agreement with the entire Reichs Cabinet, has applied to the Reichs President for the Reichstag to be dissolved in order to give the people the opportunity to create such a majority."

And so yet another round of elections in Germany happens December 7th, after prior election that just took place in May. The SPD rises by 6% points to 26%, the KPD is at 9%, the NSDAP down from 6.6% to just 3%. The republic is far from stable, but the Nazis are still just a minor splinter party.

Across the channel, the English are hardly doing any better. Here, too, a government came into office at the beginning of the year but did not survive to its end. In January 1924, Ramsay McDonald had been invited to serve as Prime Minister to form the nation's first-ever Labour government, including ten working-class cabinet members. It was seemingly the end of Eton-Oxbridge rule, the end of Conservative-Liberal dominance. From then on and until today, the Liberal Party was relegated to third place. But by

November the party is already over, MacDonald resigns. The evening edition of the *Berliner Tageblatt* from November 04 ran the headline "Today's Cabinet meeting in London":

> *MacDonald's resignation today or tomorrow.*
> *The enquiry into the Zinoviev affair.*
> *(Telegram from our correspondent.) London, 4th November.*
> *It will not be decided until the afternoon, the Parliamentary correspondent of the "Times" believes, whether the Prime Minister will present his resignation to the King today or not till tomorrow, as one of the chief duties of the Cabinet meeting to be held this afternoon will be to receive the report of the Committee appointed to inquire into the Zinoviev affair, which will probably take a good deal of time.*

What is the Zinoviev Affair? A letter leaked to the press in the middle of the election campaign claiming that elements of the Labour Party were working with the Soviet Union to increase communist influence in the UK. The letter is a forgery. Allegedly, the British domestic secret service MI5 already knew this at the time but kept quiet to help undermine the fledgling labour movement. It worked; MacDonald had to go, and the Conservatives were able to take over the government. Stanley Baldwin as the new PM restored normality to the nation in the eyes of the establishment and brought a certain Winston Churchill into his cabinet as Chancellor of the Exchequer.

Glancing across the Atlantic, Woodrow Wilson, President of the USA from 1913 to 1921, died in 1924. Wilson had already experienced crucial US historical events before ascending to the presidency: the American Civil War, the era of reconstruction in the USA, and the end of slavery. Experiences that informed his political outlook as he tried to shape the world after World War I. His main goal was to found the League of Nations, which he achieved. And this earned him the Nobel Peace Prize in 1919. But not everyone really spoke well of Wilson, certainly not the author of the following article from the *Karlsruher Tagblatt* of November 22nd, who did not want to forgive him for the Treaty of Versailles:

The monstrosity of the Versailles Dictate and the Executioners' Treaty lies precisely in the fact that those in power at the time, headed by Woodrow Wilson, who would have been far more suited to being a Methodist preacher than an authoritative head of state, played the moralist card. How did they want to take off the hypocritical mask according to which they had gone to war in perfect selflessness only for "freedom and justice" without slapping themselves in the face? The fact that they "morally" condemn us Germans does not mean that we are morally incriminated! Since we have been condemned—as the alleged instigators of the most sacrilegious of all wars, which we brought about in order to seize world domination, it is only a matter of defending ourselves against this insane accusation.

The deep resentment makes it clear that the Treaty of Versailles had by no means cleared the way for a reliable peace, but, on the contrary, was like a thorn deep in the side of the German people, causing a pain that would not subside.

This was also the year that Lenin died, leading to a momentous power struggle in Soviet Russia. The *Vorwärts* wrote on November 25, 1924, under the headline "Trotsky's Reprimands":

The sudden resurgence of the controversy over Trotsky and the possibility of his expulsion from the Party undoubtedly creates a situation from the communist point of view, the seriousness of which the Party is aware of. Even in the past party dispute it was recognized by Trotsky's opponents that since Lenin's death he was the party leader who enjoyed the most authority among the masses of the party.

…but that authority was of very little use to him. He was to lose the power struggle to succeed Lenin against Joseph Stalin. Things went downhill for Trotsky from then on. In January 1925 he lost his post as commissar, then he and his supporters were branded as dangerous Trotskyists, as "lackeys of fascism,", and in 1926 he was thrown out of the Politburo and immediately out of the CPSU, before being finally exiled to Kazakhstan in 1928.

Finally, let's take a look at the *Glocke* from December 8th, which reports in obsessive detail, about the trial in Hanover against the serial killer Fritz Haarmann. Haarmann loved young men and would regularly get into such a frenzy of pleasure during sex that he would bite his partner's Adam's apple. According to the forensic pathologists, demise came about when Fritz finally bit through the Adam's apple while choking his victim. Twenty-four boys and young men fell victim to Fritz Haarmann's bizarre behavior before he was finally caught in the summer of 1924. "Oh, believe me, I'm healthy. I only go on my tour sometimes." That's how Fritz Haarmann saw it, who considered himself innocent because it was the boys who sexually excited him and thus sealed their own fate. In any case, after the murders, Haarmann first rested a little, then made himself a strong cup of coffee and finally began to dismantle the corpse in professional manner, disposing of it little by little. The *Glocke* is indignant:

> *Below is the continuation of the Haarmann trial, which is directed against the kind of villains the world has rarely seen. In our opinion, Haarmann is undoubtedly mentally defective, but whether he was incapable of fighting off his lust for murder is another question. In any case, he is a monstrous fiend who must be excluded from human society forever, which is undoubtedly best done by execution.*

Fritz Haarmann surely would counter that this could be written only by someone who does not know the pleasure of biting into a crisp Adam's apple. The last sentence, by the way, that the safest way to exclude people from a society is by execution, is remarkably prescient. How true! How true! And that's exactly how the story ended: on April 15, 1925, Germany's most prominent executioner, Franz Friedrich Carl Gröpler, took care of the serial killer and separated his head from the rest of his body, just as Haarman had done so meticulously with so many people's heads. And just as planned, he promptly perished.

And so the last newspaper is now folded and placed back on the stack. The reading glasses are taken off, the pressure points at the bridge of the nose are briefly massaged and the glasses put back into their case. It's time to finally get changed and head to the New Year's Eve party.

1924, what kind of year was it then? At the very least a year of change. Certainties are being questioned; Generations changing. Nothing illustrates this more than the deaths of Vladimir Lenin and Woodrow Wilson. They each shaped their era. Their departure sent shockwaves around the world. And Ramsay MacDonald and Mahatma Gandhi, they represent tomorrow. Something politically new is entering the world stage. Perhaps the Atlantic flight of the LZ 126 is the event that best represents the twilight of this year 1924. The event can be interpreted as a look back as well as a look forward. It can be understood as an invoice of the cost of war just as well as a symbol of the dawn of technological modernity. Dawn and dusk at the same time. A year of unrest in the world, which always precedes upheaval to come. All in all, a fitting backdrop for what is happening in physics. The New Year's Eve party of 1924 heralds a very special year, the actual birth year of quantum mechanics.

1925, Thesis: Matrix Mechanics

January 1925, Hamburg

In January 1925, Wolfgang Pauli presented his greatest individual scientific achievement: the elaboration of what is now known as the Pauli Principle. Today, his name is primarily linked to this principle and every scientist knows it because this principle is a prerequisite to understand the structure of the periodic table. "On the connection between the closure of the electron groups in the atom and the complex structure of the spectra" appeared in the *Zeitschrift für Physik* on January 16, 1925.[1]

Considering that he had already submitted a paper to the same journal just 6 weeks earlier, one cannot help but notice that Wolfgang Pauli was a man of truly extraordinary creativity. And by comparing both works you can see how Pauli slowly and methodically pushes his way into the hidden realm of quanta. He felt committed to his new

---
[1] Pauli W. (1925)

resolution that you don't always have to be able to imagine everything. And true to this intention, he asked himself what, of all that had been written and thought about regarding the theoretical description of the quantum world in the atom, actually holds true? What is a mere idea and what is actually correct? And what he wanted to hold on to were the quantum numbers, i.e., those strange integers that were introduced to clearly determine the "states" of electrons in the atom. And all without being able or willing to explain these quantum numbers in more detail. A main quantum number n, which had to have something to do with the energies of the electrons. A quantum number k, which has to be associated with the orbital angular momentum of the electron. And a magnetic quantum number m1, which could be traced back to a component of the orbital angular momentum and belonged to the magnetic momentum of an electron. Three quantum numbers for one electron. This was still consistent with the framework of the Bohr-Sommerfeld model. But already in his letter to Landé from November 1924, he described the luminous electron in the alkali atom with four quantum numbers by assigning to it an additional separate magnetic quantum number, m2. Essentially, this assignment of a new quantum number, a fourth degree of freedom, already introduces spin, because the word "spin" stands for the intrinsic angular momentum of the electron, to which the quantum number m2 refers.

With the four quantum numbers n, k, m1, and m2 he described an electron in the atom and using this classification scheme and on the basis of the work of Edmund Clifton Stoner, Pauli was now able to formulate his famous exclusion principle. Then in January 1925, he traveled again to Ernst Back and Alfred Landé in Tübingen to test his ideas using their spectroscopic materials. The experimental results confirmed his ideas and so he sent away his legendary work in January 1925: The exclusion principle and its triumphant application to the periodic table of elements. In the original Pauline version the principle was:

There can never be two or more equivalent electrons in the atom for which the values of all quantum numbers n, k, m1, m2 match in strong fields. If there is an electron in the atom for which these quantum numbers have certain values, then this state is 'occupied'.[2]

This is the Pauli principle. The Exclusion Principle. The sentence after bluntly states:

We cannot give a more detailed reason for this rule, but it seems to present itself quite naturally.

And only with this principle can one really understand the periodic table of the elements. Because the Pauli principle ultimately answers the question of why all electrons in atoms do not behave in the same way. Rather, it is the case that—as you move up the periodic table from element to element—you add additional electrons, thereby strictly observing a certain principle for filling electron states, namely the Pauli principle. If states are already occupied, then new, higher states must be opened. In this way, shells and subshells are filled up so that when they are full, new shells and subshells are created for higher quantum numbers. It's like filling an empty bookshelf with many shelves with books from bottom to top. It is impossible to store all the books in the same place on the bottom shelf. In fact, whenever one shelf is already full, you start filling the next shelf up.

Basically, the structure of the periodic table is a representation of the filling scheme of electron states in the atoms.

---

[2] Pauli W. (1925).

And this principle is: two electrons in an atom must never have the same set of quantum numbers. No electron can be identical to another. If you want to go further, the Pauli principle explains why matter has an extension, a volume. Paul Ehrhardt commented on this in his speech on the occasion of the awarding of a prize to Pauli:

> *Take the Bohr model for a lead atom, for example. Why do only a few of the atom's 82 electrons travel along the quantum orbits close to the nucleus, but all the others travel in ever wider and wider orbits? The attraction of the 82 positive charge units of the atomic nucleus is quite powerful. Far more of the 82 electrons could therefore be drawn together on the inner quantum orbits before their mutual repulsion becomes too great. So what prevents the atom from becoming much smaller in this way? Answer: Only the Pauli Exclusion Principle: "No two electrons in the same quantum state!" So that's why atoms are so unnecessarily thick; that's why the stone, the piece of metal etc. are so voluminous!*[3]

Pauli's discovery made a great impression on the experts. Paul Ehrenfest and Arnold Sommerfeld expressed themselves outright exuberantly, and Niels Bohr was initially also "enthusiastic about the many new beauties." Finally, Heisenberg writes mockingly that Pauli "takes the dizziness to a previously unimagined, dizzying height" when he sees "individual electrons with 4 degrees of freedom." Years later, Niels Bohr revealed a little more and once again explicitly pointed out the importance of the paper by Edmund Clifton Stoner:

> *But before the Pauli principle we found out how the electrons are distributed in the atom. ... Pauli was wonderful, but there*

---

[3] Paul Ehrenfest, speech on the awarding of the Lorentz Medal to Professor Wolfgang Pauli on October 31, 1931, see Enz and von Meyenn (1988, p. 43).

is absolutely not a word that is new in the Pauli principle. That was all done by Stoner. I studied first of all the difference between the hydrogen spectrum and the other spectra. But then the other spectra had a fine structure, and in my paper that was a very bad thing, but I didn't know. And it was (probably even) not done. But Catalan had done a lot about the spectra. And that Stoner took into account, and then everything came in order. I was really only interested in how the periods start, but all the details were wrong, you see…. Pauli took the Stoner paper as a revelation, but he also did some other work on the Zeeman effect. One has the Paschen-Back effect which shows how the anomalous Zeeman effect by large fields transforms into the normal Zeeman effect in an odd manner. And that Pauli was able to calculate through and just to show how it went, and with the same idea as Stoner. That was such a fine piece of work, so it was called the Pauli principle, but one could have really called it the Stoner principle.[4]

The young physicist Ralph Kronig was the first to interpret the fourth degree of freedom of the electron as the rotation of the electron around its own axis, as its own rotation so to speak. He told Pauli about it during his visit to Tübingen in January 1925. "That's a really funny idea," commented Wolfgang Pauli, but he didn't think it corresponded to reality in any way. Kronig drops the topic and that is how the two students Uhlenbeck and Goudsmit finally achieve their somewhat dubious fame. The English physicist Thomas later commented on this event in a letter to Goudsmit:

I think that you and Uhlenbeck were very lucky that your work on the rotating electron was published and discussed before Pauli heard about it. It seems that Kronig believed in the rotating electron more than a year ago and had worked out something. The first person he showed it to was Pauli. Pauli mocked

---

[4] Bohr (1962).

the whole affair to such an extent that the first person was also the last and no one else heard about it. All this shows that the infallibility of the divine being does not extend to his self-appointed representative on earth[5]

Wolfgang Pauli: the self-proclaimed representative of the divine being. One can be sure that he would have liked this assessment.

February 1925, Berlin

February 28th at quarter past ten, Reich President Friedrich Ebert dies at the age of only 54 as a result of prolonged appendicitis. If he had undergone surgical treatment immediately at the beginning of February, the course of German history would have been different. It was certainly not a good moment for this preventable death. Hyperinflation had finally been brought under control and people generally believed that they were finally back on solid ground. And then this. Friedrich Ebert, first chancellor of the unloved republic and the most important driving force in the reconstruction of Germany. Immediately after the abdication of Wilhelm II, he was elected chancellor and the first thing he had to deal with was the November Revolution. Ebert's measures were and are controversial and the subject of much debate. Joins forces with the nationalists and conservatives and gives the Freikorps a free rein! Oh well. Out of fear of the left, of the Spartacists. Probably a mistake, given that it was the same nationalists who were attacking the republic from the start. This decision weighed on the leadership of the SPD for many years. It had betrayed the working class and essentially could never rid itself of this

---

[5] Thomas, L. H. in a letter to Samuel Goudsmit dated March 15, 1926.

error. The situation was exacerbated by the murders of the prominent left-wing leaders Karl Liebknecht and Rosa Luxemburg, most likely carried out with the approval of the SPD leadership.

Just a month later, the election for president took place and it became clear that Ebert's death had significantly weakened the Weimar Republic. The Republic's candidate failed and Paul von Hindenburg became the new president, a man of the nationalist right, who formally supported the Weimar Constitution but was far from a convinced democrat. This led to years of political instability. From 1924 to 1929, only three out of seven governments had a majority in the Reichstag. The Weimar Republic languished, with a conservative revolution threatening from the right and a proletarian revolution from the left. And since the Reichstag was weak, political life moved to the streets. The time of fighting alliances had begun.

March 1925, Hamburg

Physics is once again very muddled at the moment. In any case, it's far too difficult for me, and I wish I were a film comedian or something similar and had never heard of physics![6]

You read this sentence by Pauli again and again when it comes to the last months of pregnancy shortly before the birth of quantum mechanics. He sums up the feeling prevailing among physicists at the time on the eve of the events on the island of Heligoland. Quantum physics was really a rather messy situation in the spring of 1925. Everyone was muddling around with it. Everyone knows that the Bohr-Sommerfeld model has served its purpose and must be

---

[6] Klein and Toomer (1979, p. 214).

replaced. Everyone knows that a radical new beginning is needed and that great discoveries are yet to come, but have to come before too long. The *Leipziger Tageblatt* of September 27, 1924, expressed this in a wonderfully old-fashioned way in a report from the Natural Scientists and Medical Congress in Innsbruck:

> *A series of lectures was dedicated to the latest results of atomic and molecular research. The most important speaker was Professor Dr Sommerfeld-München, who dealt with the fundamentals of quantum theory and Bohr's atomic model. Even if quantum theory does not yet appear to be ready for a purely deductive conception, the ideal form to which it will presumably aspire can already be surmised. Regardless of all subsequent discoveries, quantum theory and the doctrine of Bohr's atomic models will always remain an inalienable possession of physics.*

Signed by Dr. H., who senses an ideal form—that only he can imagine. Somehow, Bohr's atomic model "as inalienable property"—how should one sell this property, dear Dr. H., no one will buy it!—gives no clue as to how things could proceed. Everyone knows that the time has come for a new, grand theory. But no one has any idea where it's supposed to come from. Already at the end of 1924 the great Einstein had sobered up:

> *My attempts to give the quanta a tangible form have failed again and again, but I'm far from giving up hope.*[7]

> *We need a higher form of order more urgently than ever. When will the redeeming thought be given to us?*[8]

---

[7] Letter from Einstein to Born dated April 29, 1924 (Born M., Einstein-Bohr Briefwechsel 1916–1955. third Edition, 2005, p. 141).
[8] Fischer (2022, p. 130).

He said this in anticipation of envying those lucky ones who had not yet been found:

Happy are those who are allowed to experience and see the higher form![9]

Yes, but who might that be? Who will be allowed to see the higher form? No one knows. No one? Wrong. Niels Bohr thinks he knows who will manage. He has chosen him and put him on the scent. As early as April 1924, while hiking together through Denmark, he put the clever Heisenberg on the right track, prepped him like a hunting dog, motivated him, programmed him, and awakened his ambition. This man can do it, he will find the needed idea. Our savior! At least that is Nils Bohr's conviction. At the end of that hike he tells one of his students:

Now everything is in Heisenberg's hands—to find a way out of the difficulties.[10]

Max Born is also very impressed by Heisenberg, the child prodigy in whom Niels Bohr so believes:

He looked like a farm boy, with short blonde hair, clear, bright eyes and a charming facial expression... His incredible speed and accuracy of perception enabled him to constantly accomplish an enormous amount of work without much effort.[11]

And elsewhere Born writes:

I grew very fond of Heisenberg; he is liked and appreciated by all of us. His talent is outrageous, but even more appealing is his

---

[9] Ibid.
[10] Hürter (2021, p. 126).
[11] Born M., Mein Leben (1978, p. 291).

nice, humble nature, his good mood, his zeal and his enthusiasm.[12]

So, this is what the coming messiah of quantum mechanics looks like: A simple farm boy, nice, humble, happy, and enthusiastic. Hard to believe.

March 1925, Bristol

On March 9, 1925, Paul Dirac received a letter from his aunt Nell. His brother Felix has taken his own life. The small Dirac family is in utter shock. In times past Paul had been very close to his brother Felix. But that was long before his university days, which had introduced a suffocating rivalry between the brothers. Felix's gnawing jealousy of his brilliant younger brother had poisoned their relationship. At the time of Felix's death, they were no longer speaking. Paul Dirac had essentially no contact with his brother and Felix's problems were completely unknown to him and the rest of the family. Felix had actually wanted to pursue a medical profession, but his father wouldn't allow it for fear of financial risk. Instead, he insisted that Felix had to study engineering, a subject he neither liked nor was good at. At the beginning of 1925, he was living in Birmingham, operating testing machines in a factory for a pittance. In the evenings he volunteered for the Ambulance Corps and dreamed of the life forbidden by his father.

For Paul, on the other hand, things were going brilliantly. At only 22 years old, he was widely seen as a remarkably capable graduate student; even for Cambridge. And it had by no means been certain that he would even make it there. His parents would not have had the money for such a

---

[12] Letter from Born to Arnold Sommerfeld dated January 5, 1923.

university so Paul had gotten in solely based on his own achievements. He was all too aware of his parents' money worries; there was not a letter from his mother in which she did not complain either about their finances or his father's stinginess. In his work, Paul was at the same time satisfied and dissatisfied. Although he had already written several papers, none included a "really original idea." And that was the only thing his ambitions and dreams were focused on: a "really original idea," so good that he would be recognized internationally. His room in Cambridge was very simple and unadorned, which actually suited his very reserved and modest personality quite well. Thus, he was still a fairly typical student. But that would change instantly with a truly original idea. He harbored no doubt about that.

In her letter, Aunt Nell actually did not have much to report. The details only came to light later on. Felix was found on March sixth, under a holly bush on the edge of a field, 3 km south of the small town of Much Wenlock in Shropshire. There he lay, elegantly dressed in a suit with a bow tie. The wrench is still in his trouser pockets and the bicycle clamps on his trousers. But where was the bike? Next to the body is an open glass bottle. Poison. Nothing else. No letter, no message. Nobody knew anything. The lonely corpse of a lonely person. In January, Felix had suddenly dropped out of his everyday life. Gave up his job, but, according to his landlady, still left his apartment every morning and returned in the evening as if he was still going to work. Where did he go in these last few months? He probably lived off his savings, as Paul Dirac later reported:

What he did during those last three months, no one was able to find out...He continued to pay his rent regularly. He just withdrew his savings, and when they were gone he killed himself.[13]

---

[13] Dirac P. (1962).

Their parents, Charles and Florence, were what you could call "emotionally distant." So much that Paul was seriously surprised that his parents showed feelings of grief after his brother's death:

My parents were terribly distressed. I didn't know they cared so much…I never knew that parents ought to care for their children.[14]

That wasn't meant ironically at all. He was completely serious. Until he was 22 years, Paul Dirac was not aware that his parents had feelings of loving care for him and his brother. Doesn't that tell us everything about his childhood? His father was a Swiss citizen and came to England as a young man to teach languages. He had moved to Bristol to take up a position as a modern language teacher at Merchant Venturers' School, one of the better schools in the area, which focused more on technical subjects. Paul's early life was dominated by his authoritarian and tyrannical father. This could go some way to explain Dirac's famous reserve and his calm, taciturn personality. When Paul and Felix were young, his father had insisted that the children only speak French to him. The mother, however, spoke English. Dirac remembers:

My father made the rule that I should only talk to him in French. He thought it would be good for me to learn French in that way. Since I found that I couldn't express myself in French, it was better for me to stay silent than to talk in English. So I became very silent at that time.[15]

And not just "at that time." He was actually silent for the rest of his life; he was even famous for it. Years later, his

---

[14] Farmelo (2009, p. 79).
[15] Dirac P. (1962).

colleagues at Cambridge searched for the smallest imaginable number of words that a linguistic being could theoretically utter in the company of others. They called this unit "a Dirac," or about one word per hour.

April 1925, Göttingen

Max Born was finally back in Göttingen. He had spent the first 2 weeks of March in Copenhagen with Heisenberg and Bohr and then went skiing in Switzerland for 10 days, in part because he hoped that the fresh mountain air would be good for his asthma. Pascual Jordan, his assistant since January, made it very clear on his return how eagerly he had been waiting for him. Jordan was absolutely sure that he and Born were backing the right horse. And Born, in turn, was sure that he had gotten a good deal with Jordan. It was unusual at the time for such a young person, and so soon after receiving his doctorate, to become an assistant to a professor of the caliber of Max Born. But Born knew what he was doing. He once said of Jordan that he is able to think much faster and more confidently than Born himself. Jordan was on his way to becoming a great mathematician, a mathematician like Paul Dirac would become. Even Heisenberg, who was widely admired for his abilities, readily admitted in a small group that Dirac and Jordan were far better mathematicians than he.

Jordan had completed his doctorate under Born the prior year and before that had been an assistant to the famous mathematician Richard Courant. The time with Courant had been important for Jordan. Typing up Courant's lecture notes on linear algebra familiarized him with the mathematics that would become so immensely useful in his later work with Born. Namely matrix algebra. He also supported

James Franck in writing a book and worked on his own doctoral thesis. An incredible workload for the young mathematician. And in January 1925 the time had come: finally, a permanent position with Professor Born at the University of Göttingen.

Unfortunately, Pascual Jordan was an unusually shy person with a severe stutter. He stuttered so badly that "it was hard to bear," said Robert Oppenheimer. And in the 1920s, stuttering was still seen as a sign of reduced intelligence, an almost insurmountable obstacle to a scientific career. Max Born, who was justifiably worried about his student's future, even paid for him to see a psychotherapist in an attempt to cure his illness—without success. Indeed, Jordan's stammer affected his career in many ways and contributed to his never being offered a truly prestigious position. Although he had demonstrated his outstanding intellectual and mathematical abilities like no other, he was rarely asked to give public lectures or take part in international lecture tours, which was absolutely essential for the professional advancement of an academic at the time.

But here and now, as Max Born's assistant, he didn't need to be in the public eye. Here he could be cocooned in the refuge of the famous Max Born. And what about Max Born? Max Born was 42 years old and had now been in Göttingen for 4 years. He was born and grew up in Breslau, where he also began to study mathematics. As a physicist, he was always a bit awkward. He once said that he lacked physical intuition. But he was an excellent and knowledgeable mathematician. And above all, he had a knack for people. He had the ability to win over and retain people who were significantly more talented than himself, to wit the group of his assistants: Wolfgang Pauli, Werner Heisenberg, Friedrich Hund, and Pascual Jordan.

So now it was April, and spring was slowly coming to Göttingen. Pascual Jordan and Max Born dedicated the spring of 1925 purely to quantum theory. The pair wasted no time and already in April, they discovered a relationship between the transition probabilities between atomic energy levels and the observed intensities of the spectral lines, a relationship of much use later on for Heisenberg. And finally, in their work "On a Quantum Theory of Aperiodic Processes," they highlighted again and again what Wolfgang Pauli was never tired of emphasizing: please create theories only of things that are actually observable!

A principle of great breadth and fruitfulness states that only those quantities that can in principle be observed and determined enter into the true laws of nature.[16]

This insistence on observability was by no means new in physics. Einstein had already based his theory of relativity on the realization that it is impossible to measure the absolute simultaneity of two events taking place in two different places, and Einstein was the reference for everything in those days. Alfred Kerr once wrote in the *Berliner Tageblatt*: "Everything is Einstein, only Einstein is not Einstein." The cult of Einstein must have been unimaginable. Born was also a friend and devout disciple of Einstein and essentially only repeated Einstein's thoughts: theories only about observable phenomena, please! And finally, in their work, Born and Jordan put forward the hypothesis that these very transition probabilities could potentially become centrally important for a new quantum theory—and Heisenberg later made this almost prophetic announcement come true. All of these early discussions in Göttingen—transition amplitudes, intensity of spectral

---

[16] Born and Jordan, Zur Quantentheorie aperiodischer Vorgänge (192, p. 493)

lines, and observability—had a great influence on Heisenberg, who was now back from Copenhagen and back at the side of Max Born.

April 1925, New Jersey

In the laboratory of Dr. Clinton Joseph Davisson on the spacious premises of the recently founded Bell Labs in Murray Hill, New Jersey, a bottle of liquid air explodes. Particularly impacted is the valuable high-vacuum tube in the experimental setup of Davisson's two assistants, Kunsman and Germer. It's completely ruined. Air has entered the tube and created a thick oxide layer on the metal surface of the platinum target inside the tube. Clinton Davisson is understandably angry because the electrode was expensive. Kunsman and Germer better see to it that the oxide layer gets removed from the electrode. This can only be achieved through long and repeated heating at various high temperatures, sometimes in a hydrogen atmosphere and sometimes in a vacuum. What was this kind of experimental setup actually for? An electron beam from an electron gun was directed at a metal surface in the high vacuum of the tube in order to examine how the electrons scattered on this surface. The electron detector was mounted on an arc-shaped guide so that the scattered electrons could be measured in a circular arc, i.e., their angular dependence could be determined.

And when the valuable metal surface was finally cleaned again and the measuring apparatus was restored, the experiments could resume. But, lo and behold, the electrons suddenly showed a completely different angular distribution than in the weeks before the mishap. Now what did *that* mean? The original sample had previously been polycrystalline. And the repeated heating during the repair had inadvertently caused the many small crystallites to melt and rearrange themselves into one large crystal. A clean single crystal with the rows of platinum atoms nicely arranged in a straight and periodic fashion on its surface. Yes, the unexpected results had to have something to do with that. Clinton Davisson sent the results, which he found completely incomprehensible, to Max Born and James Franck in Göttingen. Max Born notes:

By a strange coincidence, just at this time a letter arrived from the American physicist Davisson, who had obtained incomprehensible results from the reflection of electrons on metal surfaces and documented these with curves and tables. When I discussed the letter with Franck, we came up with the idea that the strange maxima in Davisson's curves could perhaps be explained by the diffraction of the electronic waves of matter in the crystal lattice, and a crude calculation using de Broglie's formulas gave the correct magnitude of the wavelength. We handed over the development of this idea to our student Elsasser, who had initially worked experimentally with Franck but wanted to switch to theory.[17]

May 1925, Cambridge

Niels Bohr finally speaks in Cambridge! The great Niels Bohr. Here in Cambridge, he had given birth to his large atomic model as a postdoc with Ernest Rutherford, and here in Cambridge he had become famous. But he had made himself scarce since winning the Nobel Prize in 1922. But now finally: in the beautiful month of May.

Niels Bohr: The man was already a legend during his lifetime, the world's leading authority in the field of quantum

---

[17] Born M., Einstein-Born Briefwechsel 1916–1955 (third edition, 2005, p. 146).

theory. Pretty average as a mathematician, but totally unique as a physicist. How can one, who couldn't even begin to hold a candle to Einstein, Schrödinger, and Heisenberg mathematically, become the father of such a mathematically sophisticated idea as quantum physics, inspiring geniuses like Pauli, Heisenberg, Jordan, and Dirac? The answer can be found in a single sentence by Friedrich Hund:

> *Bohr wasn't afraid to talk about things that couldn't actually be talked about yet.*[18]

All other physicists simply did not dare to talk about the new and completely unknown. They didn't have the language for it. Especially in mathematics. They simply couldn't bear what was incomplete. They wanted to lecture about what was complete and already fully understood and were ashamed of the style of helplessly searching and the uncertain language necessary to discuss something not yet understood. But Niels Bohr simply wasn't embarrassed. Not for his searching language, not for his uncertainty, not for searching in general. Precisely this searching made him who he was, possibly the most passionate seeker the world knew. And by thinking out loud to himself and out loud in front of everyone else, by unabashedly searching not only for words but also for new truths, he captivated people, especially the very young physicists. And he transformed science from the proclamation of finished conclusions to a science of collaborative seeking and discovering. "Bohr wasn't afraid to talk about things that you couldn't actually be talked about yet." That somehow became the new approach in international physics: not to be ashamed of your

---

[18] Dr. Konrad Lindner, "Friedrich Hund über die Bohrfestspiele 1922," https://www.leipzig-lese.de/persoenlichkeiten/h/hund-friedrich/friedrich-hund-ueber-die-bohrfestspiele-von-1922

search and of your own, initially rather simple language. Einstein sums it up when he says about Bohr:

He expresses his opinion like someone who is constantly searching and not like someone who believes he is in possession of the ultimate truth.[19]

And now this singular Niels Bohr is coming to Cambridge. Paul Dirac is totally thrilled. Of course, Dirac has already dealt with Bohr's atomic theory, studied Fowler's lectures and Louis de Broglie's doctoral thesis, as well as the classic textbook "Atomic Structure and Spectral Lines" by Arnold Sommerfeld, the book that Pauli and Heisenberg called their "Bible." But naturally Dirac was particularly interested in any piece of mathematics that could bring clarity to physics. He was enthusiastic about the theory of relativity and Maxwell's equations because in just a few equations they present a fundamental theory that is able to summarize a huge range of phenomena and experiments in utmost succinctness. Fundamental formulas of complete simplicity that absorb a lot of things and summarize them in the most beautiful unity: to find something like that and give it to the world was what Dirac dreamed of. That was the "truly original idea" he sought so badly.

Bohr's May lecture is indeed extremely stimulating for the young Dirac. It is becoming clear: Spring has arrived for quantum theory as well! The current thoughts regarding atomic theory are only a beginning, says Bohr, and predicts that his own model will very soon be completely outdated, now that a young generation of physicists is entering the field to sort things out. That was music to the ears of Paul Dirac. That sounded like an invitation. There must be a really original idea for him somewhere!

---

[19] Hürter (2021, p. 122).

June 1925, Berlin

On June 10, 1925, the dancer Anita Berber turns 26 years old. Together with Josephine Baker, she was probably the most famous dancer of the 1920s. Her dance was of a unique kind, spontaneous, provocative, and uninhibited. She often danced naked and was not bothered by the commotion that ensued. "Dances of Vice, Horror and Ecstasy" was the name of one of her stage programs, incidentally, completely sold out. Anita Berber—the scandalous femme fatale of her time, also led a completely uninhibited private life characterized by alcohol and drug addiction. A bottle of cognac a day. At a minimum. Sometimes performances were canceled because her excessive cocaine consumption no longer allowed coordinated movements. But the world of men was at her feet. And since she, the great idol of the Weimar era, didn't care about anything, men could even buy her body. Marta Dix claimed that it cost 200 marks. She was lavish, decadent, oblivious to the world, wild, and above all: uninhibited. And thus became a symbol of the Weimar Republic, the end of which she would never see.

In 1925, she stood completely naked for Otto Dix, who painted her so emaciated, so old, so worn-out, though she surely didn't look this bad despite her excesses. Rendered completely in shades of red. Only the dancer's face is bright white, in brutal contrast to the redness of the dress and the background. The makeup on the face, the clawed hand, the chiseled body. Anita Berber models in the nude and yet is painted in a red dress. Why would she model for such a caricature? But the vitality, the excess, the decadence of the Berlin dancer jumps out at you. This is the style of the "Neue Sachlichkeit," of which Otto Dix is the high-priest.

The horrors of the First World War shaped many German artists of the 1920s. Otto Dix, who fought and survived the

Battle of the Somme and was awarded the Iron Cross for his bravery, focused his artistic work on the Weimar Republic's attempt to get over the horrific experiences of the World War. He painted wounded and disabled veterans, homeless people on Berlin streets, and depictions of life in the trenches. The war painting "The Trench" so disturbed its viewers that the museum director of the Wallraf-Richartz Museum had to resign. With the picture of Anita Berber, Otto Dix once again succeeds in distinguishing himself as a brutally direct and ruthlessly honest contemporary witness. A well-known photograph of Werner Heisenberg, who would finally begin the real history of quantum mechanics in distant Heligoland in the same month, simply has to be placed next to Dix's painting of Anita Berber. Here the red, sinful dancer, there the well-behaved, honest young man with hat, tie, and suit. Times of contradiction, of uncertainty, of opposites. But perhaps that is just a superficial observation, and these are not opposites at all: perhaps one expresses her new attitude to life in uninhibited, boundary-pushing dance, while the other succumbs to the same feeling by overcoming boundaries in physical thinking. Werner Heisenberg and Anita Berber, partners in spirit? A couple freed from the confines of the nineteenth century. In any case, they are both very much children of the twenties!

For other artists the 1920s were a time of reflection, i.e., of thinking about the past. "It is no longer pain, but contemplation," Käthe Kollwitz once wrote. In some way, the war had touched them all, including the grieving Kollwitz, who lost a son in the First World War, then a grandson and her husband in the second. The sculpture "Grieving Parents," created between 1914 and 1932, is dedicated to the fallen son. Kollwitz, a member of the German Artists' Association and the Berlin Secession, was deeply horrified by the growing militarism of the 1920s. Her work has a strong anti-war message and depicts the oppressive life of the German working class. She made countless woodcuts and prints that showed death in times of war in a variety of ways. She also dedicated a woodcut to Karl Liebknecht after his murder. According to Kollwitz, art has the task of depicting the social conditions of its time. Her most beautiful work is probably the Pieta sculpture from the 1930s, a replica of which is now in the Neue Wache in Berlin and which shows a grieving mother cradling her dying son. This is sadness cast in bronze, a feeling that also colored those years alongside the naked, wild excesses of Anita Berber.

June 1925, Heligoland

On June 7th, Werner Heisenberg finally had it. His hay fever is absolutely unbearable this year. He looks like he's been beaten up. Nose and sinuses hurt, his brain is completely foggy. He presents such a pitiful picture that Max Born lets him escape to the island of Heligoland without hesitation. Because not a tree, not a bush, and barely a flower is found here. Add a strong, fresh sea breeze. In short, no pollen. A life raft for all those who suffer from hay fever! Werner Heisenberg stays on the island for 10 days, takes long walks, goes swimming regularly, memorizes Goethe's poems from the West-East Divan and otherwise sits on the small balcony of his vacation rental to write one of the arguably most important scientific works of the twentieth century. Which, of course, at this point he does not know or even suspect. All he knows is that he wants to make it happen. So he sits with his calculations, ponders them, obsesses, and thinks about nothing else: day and night. Heisenberg on Heligoland: the view toward the open sea, plus the fresh wind and Heisenberg begins to think broadly and clearly. This is where quantum mechanics is born.

But quantum mechanics is not born with a single contraction. Not at all. It took several such contractions for this great theory to be born. The actual birthing lasts almost two years. Plus a full year to figure out with others what a strange being has been brought into the world. 1925 is notable for the works of the trio from Göttingen, Werner Heisenberg, Pascual Jordan, and Max Born, who developed matrix mechanics as one of two possible formulations of quantum mechanics. Then 1926 adds the contribution of Schrödinger, who discovered wave mechanics equivalent to matrix mechanics as the second formulation of quantum mechanics. Interspersed is the work of Paul Dirac and Pascual Jordan. And finally the heavy lift of interpreting this theory: from Max Born's interpretation of Schrödinger's wave, via Heisenberg's uncertainty principle to Bohr's Copenhagen interpretation.

So let's go to 1925 and the three Göttingen articles. The first article with Heisenberg as the sole author reached the *Zeitschrift für Physik* in July 1925 and was largely written on the balcony of Heisenberg's apartment in Heligoland. The second article in the same journal is by Max Born and Pascual Jordan, builds directly on the first article, and was submitted in September. And then there is a third article, the so-called *three-man work* by Heisenberg, Jordan, and Born. Heisenberg received a Nobel Prize for his contribution early on, Born only much later and Pascual Jordan not at all, although when studying this whole story you get the impression that it was actually only through the mathematical skills of Pascual Jordan that Heisenberg's ideas became a real theory. In any case, the second paper by Born and Jordan is much clearer and more understandable than the first paper by Heisenberg, which, while widely hailed as the decisive breakthrough, is at the same time known as a paper that is extremely difficult to understand. Steven Weinberg, the Nobel Prize-winning physicist, puts it this way:

I have tried several times to read the paper that Heisenberg wrote on returning from Heligoland, and, although I think I understand quantum mechanics, I have never understood Heisenberg's motivation for the mathematical steps in his paper.... Heisenberg's paper was pure magic.[20]

The fact that the other two Göttingen papers follow so closely and that authorities like Born and Jordan, but also Paul Dirac, immediately and enthusiastically jump on this bandwagon, proves that these contemporaries really appreciated Heisenberg's paper as a groundbreaking idea. It contains a deep and completely new idea of physics that many people later saw as the saving grace. Born and Jordan in the second Göttingen paper explain why they consider Heisenberg's Heligoland paper to be the breakthrough:

The approaches to new kinematics and mechanics that correspond to the basic requirements of quantum theory, which Heisenberg recently communicated in this journal, seem to us to be of great importance. They constitute an attempt to do justice to the new facts—instead of through a more or less artificial and forced adaptation of the old, familiar concepts—by creating a new, truly suitable system of concepts.[21]

So that's what's special: the new system of concepts. Born and Jordan don't want to call it a theory yet, it is more like an approach to a theory. But the paper now contains a decisive hint as to how one can get to the actual theory. First of all, Heisenberg explains in his article what he learned

---

[20] Weinberg (1992, p. 53).
[21] Born and Jordan, Zur Quantenmechanik (1925).

from Wolfgang Pauli and Max Born: if you want to make a theory, you can only try to describe theoretically those things that are actually observable, at least in principle. It is completely pointless to bother with describing things that no one will ever be able to observe. To wit: the orbit of an electron in an atom is just such a thing that no one will ever be able to observe. An electron orbit is not observable and therefore should not appear in a theory. And even if we can all imagine how an electron moves in circular orbits around an atomic nucleus like a planet around its sun, we must not succumb to the temptation to construct a theory based on this imagination. And so the theory of electrons in the atom simply has to do without the concept of orbit. This is certainly Heisenberg's first radical step. A real and final farewell to the concept of an electron orbit. But what then takes the place of an orbit? Answer: what you can observe. Heisenberg writes about this in his paper:

We must remember that in quantum theory it was not possible to assign the electron a point in space as a function of time using observable quantities. However, in quantum theory the electron can also be assigned a radiation. This radiation is described firstly by the frequencies. In addition to the frequencies, the amplitudes are also necessary to describe the radiation.[22]

So the radiation of an electron as an actual and new descriptive quantity takes the place of an orbit. You scratch your head. What did we gain from this distinction? Be patient! Now Heisenberg takes advantage of the correspondence principle, as he learned it from Niels Bohr. Every mathematical relationship in quantum mechanics needs a counterpart in classical physics. Heisenberg adheres to this principle almost slavishly: every relationship in his paper

---

[22] Heisenberg W., Über quantentheoretische Umdeutung kinematischer und mechanischer Beziehungen (1925).

appears twice, once as an expression of classical physics and a second time as an assumed quantum mechanical expression, ideally constructed parallel to it. And at first it is always about expressions that describe the radiation of an electron. But then Heisenberg suddenly reaches a point where the classical expression can no longer be placed alongside the quantum mechanical one. In the classical description, he adds up expressions and the sum leads back to the orbital concept. In the quantum mechanical description, however, he cannot perform this calculation. It comes to a standstill. It simply doesn't work and so he realizes that it was appropriate to abandon the concept of the orbit. He writes:

> *Such a unification of the corresponding quantum theoretical quantities does not seem possible without arbitrariness and therefore does not make sense; However, the entirety of the amplitudes can be viewed as a representative of the quantity $x(t)$.*[23]

And here lies the second radical step in Heisenberg's work and the answer to the question of which descriptive quantity will now replace the orbit $x(t)$. If I sum up something in the classical description and thus arrive at the concept of orbit, but cannot do the same in the quantum mechanical description, then I simply refrain from summing up and take the original, non-summable expressions and use them in place of the orbit concept. These new quantities can be named: they are the transition amplitudes between atomic energy levels. Now the name isn't that important. Of importance is that it is no longer a single quantity (the orbit of the electron in classical physics), but—precisely because it was not possible to sum them up—a large number

---

[23] Heisenberg W., Über quantentheoretische Umdeutung kinematischer und mechanischer Beziehungen (1925, p. 882).

of simultaneous quantities, indeed an infinitely large table of numbers, a so-called matrix.

In other words: Heisenberg invents a mechanic to describe electrons using a quantum theory that is based on matrices of transition amplitudes as a descriptive quantity instead of the electron orbit. That is the idea!! The new theory is to rest on these new quantities, on these pillars! And these amplitudes evolve in later papers into the transition probabilities between states and are therefore observable. Just like the frequencies. Heisenberg does not yet call the new descriptive quantity a matrix because at this point in time he does not know the matrix concept or the associated mathematics. But he nonetheless discovers that if you multiply two of these matrices, the result depends on the order. A times B is not B times A. Matrix multiplication is not commutative, which is disconcerting to him, but later tremendously fascinating to people like Paul Dirac. He ends his magical paper by calculating a small example, the anharmonic oscillator, to demonstrate his new quantum mechanics.

The paper itself was probably created in Göttingen, but the actual calculations were made in Heligoland. Much has been written about Heisenberg's stay here; some of which sounds as if an American screenwriter wanted to write something for Hollywood. But let's let someone else tell it: Tobias Hürter, in his very beautiful book "Das Zeitalter der Unschärfe":

His hay fever has subsided and now the new, inner fever is robbing him of sleep. But this time he is not disheartened. One night at around three o'clock, the result appears on the paper in front of him. The excitement prevents Heisenberg from falling asleep. Nearly intoxicated with arousal, he ventures into the twilight of dawn and hikes to the southern tip of the island to climb the "Monk", a 55 meter high, rugged rock tower.

> *Heisenberg reaches the top unscathed, sees the sun slowly rise and descends safely. It is the most precarious climb in the history of physics. Heisenberg's head is the only place where quantum mechanics existed. If he had crashed, it too would have been irretrievably lost.*[24]

By the way, that rock tower on Heligoland collapsed when the British tried to blow up the island in 1947. But that's just an aside.

July 1925, Göttingen

On July 23rd, Max Born wrote down work assignments for Jordan and his other assistants in his diary and also gave himself an assignment with the keyword "Heisenberg's quantum mechanics." He wanted to study the article that Heisenberg had started on Heligoland and submitted to him as one of the editors of the *Zeitschrift für Physik* on July 10th. Heisenberg had told him that he didn't really know what he had found on the island, he was unsure, and could Born decide whether it should be sent to the Journal? He then disappeared to Cambridge, where he had received an invitation to speak.

But the paper had an effect on Max Born that was difficult to describe; it literally worked on him and he couldn't sleep for several nights because he was so excited. Then he finally understood: the strange calculation that Heisenberg had unknowingly written down was "nothing but the well-known matrix calculation." He easily could have saved himself the sleepless nights and reached this insight more quickly if only he had asked his assistant Pascual Jordan. After all, the year before Jordan had typed up the lecture

---

[24] Hürter (2021, p. 141–142).

notes on linear algebra for Professor Richard Courant and discussed them intensively with Courant, so he had matrix algebra practically at his typewriting fingertips. On July 19, Born went to Hanover for the meeting of the German Physical Society, where he met Wolfgang Pauli. Wouldn't he, Pauli, want to work on Heisenberg's brilliant idea together with him? The answer was typical Pauli, arrogant and brutally direct:

> *Yes, I know you are a supporter of such lengthy and complicated formalisms. You will destroy Heisenberg's physical ideas with your useless mathematics.*[25]

Today one might think: pretty dumb, dear Pauli! First of all, Born's eventual mathematically oriented paper is by no means useless, but on the contrary extremely helpful in understanding Heisenberg's ideas. And secondly, you, Pauli, just let a once-in-a-century opportunity slip through your fingers, namely to go down in history yourself as the founder of quantum mechanics. Well, pride rarely if ever pays off.

In any case, Born likely was not particularly pleased with Pauli's answer. And so now Pascual Jordan comes into play. He had—thank God!—overheard the conversation between the two. And when the rude Pauli was gone, he pointed out to Max Born in his shy, introverted way that he definitely knew something about matrices and that Born could work it out with him. Born agreed and wouldn't regret it. Quite the opposite! After the diary entry from July 23rd, Born actually worked on Heisenberg's famous paper for four full days without interruption and then finally sent it off to the *Zeitschrift für Physik*.

And so what is probably the most important paper in our history sets off on its journey. It marks a beginning. At the

---

[25] Born M., *Mein Leben* (1978, p. 300).

same time, a completely different book also marks a beginning. A book that sees the light of day almost at the same time, on July 18th, and sets off on a journey that is at least as long. It is an incoherent series of angry rants and dull ravings poured onto endless pages by a narcissistic control freak. Some of it is autobiographical, but mostly agitation and propaganda. Regardless, it forms the foundation for the darkest era in Europe's history. *Mein Kampf*, written by Adolf Hitler while serving a nine-month prison sentence after the Beer Hall Putsch was published on July 18, 1925.

Hitler describes his perceived struggles and insults from various groups. He complains in a whining fashion about his outsider status and indulges in fantasies of power and control. Just as in Oswald Spengler's *The Decline of the West*, Hitler sees the threat of decline of Western civilization. His treatise reveals that the core of his ideology is formed by conspiracy theories. After Hitler's power grab sales of the book skyrocketed, eventually reaching five million copies by 1939, from pocket editions for soldiers at the front to wedding editions for newly married couples. Hitler himself was ambivalent about it in later years, but there is no doubt that it acted as a blueprint and served as an early warning of what the Nazis would do with, and in, Germany.

July 1925, Dayton

The young man's name was John Thomas Scopes. He was 24 years old and a high school football coach in Dayton, a sleepy town in the US state of Tennessee. Now and then he also had duty as a substitute teacher, for example, on the very day that the biology teacher at his school was sick. So John Thomas Scopes found himself teaching the children entrusted to him biology and introduced them to Darwin's theory of evolution. But that was a problem in Tennessee. The Butler Act had been passed a few months earlier, criminalizing the presentation of the theory of evolution in school lessons and instead recommended for school curricula the version of creation presented in the Book of Genesis. In response, the ACLU, the American Civil Liberties Union, decided to test the new law in court. The ACLU reckoned there had to be a teacher somewhere in Tennessee who could be persuaded, with adequate compensation, to teach the theory of evolution in his class and thus knowingly violate the Butler Act.

A group of shrewd Dayton businessmen immediately shouted "Here!" hoping that such a story would be good publicity for the city and equally good business for its residents. So they looked and found John Thomas Scopes, who wanted to earn a little extra money and so willingly heeded the call of the ACLU. And that is how John Thomas Scopes ended up in court. His defense was carried out by a star lawyer from Chicago, and the prosecution was represented by a former US Secretary of State. The trial began on July 10, 1925, and the whole world watched: it was the showdown between supporters and opponents of the theory of evolution. And the Dayton business community's bet paid off. There was an indescribable media frenzy. No fewer than 5000 tourists and journalists from all over the world came to Dayton while the rest of the world could follow the radio broadcast. A phenomenal spectacle! And the little town of Dayton is at the center of this "trial of the century."

Far away in Göttingen Pascual Jordan followed this trial as well. He, who was to become one of the great scientists of his time, was by no means solely on the side of science. Deeply religious himself, he could understand very well why people in Tennessee struggled so much with Charles Darwin and his teachings. At the age of 12 years, Pascual

Jordan also had an internal struggle that was just as fierce as the fight in the Scopes trial. It was the biggest crisis of his childhood. His interest in science had just been awakened, and now this Darwin! With red cheeks, he discussed how Darwin's theory of evolution could be reconciled with his faith and the literal truths of the Bible. "It seemed to me a tormenting nuisance,"[26] he said later. After intensive discussions with a teacher, he finally came to the conclusion that one could accept the theory of evolution without having to become an atheist.

Pascual Jordan is certainly one of the most enigmatic characters in this story. On the one hand, a brilliant mathematician who rose in the scientific world with astonishing speed. On the other hand, someone who struggled not only with Darwin's theory of evolution but also with modernity as a whole. Initially just an avowed conservative, he quickly became a nationalist conservative in the turbulent Weimar Republic and, when the time came, an enthusiastic convert to National Socialism. Whether it was the theory of evolution and his beliefs, or his close collaboration with his Jewish boss Max Born and his enthusiasm for the Nazis, Pascual Jordan seemed to live in two fundamentally different worlds at the same time. He was somehow able to completely separate his religious fundamentalism, his nationalism, and his anti-Semitic beliefs from his other life as a scientist. Strange!

> *There are not many physicists in whose biography the contradictions of human existence, the closeness of glorious scientific achievements and the disturbing human weaknesses in the face of the great catastrophe of the twentieth century are so clearly reflected as in the personality of Pascual Jordan.*[27]

---

[26] Jordan P., Philosophie in Selbstdarstellugen I (1975, p. 197).
[27] Schroer (2003, p. 1).

This is how the historian Bert Schroer put it. But internal contradictions were generally a sign of the 1920s, something that was present throughout the entire period. Contradictions and inner conflicts that can hardly be imagined today.

This also came to light in, and with, the trial of John Thomas Scopes: two worlds that attacked each other and tried to crush each other. The arguments took many turns, covering the story of Adam and Eve, the personal freedoms of teachers, and the intricacies of creationist theology. Both sides argued their case passionately and with such vehemence that the judge had to intervene several times to restore order in the court. Although the trial took a long time and led to a lot of public discussion, the jury did not need such discussions and completed their task quite quickly. It took them all of 9 minutes to reach an agreement among themselves and find Mr. Scopes guilty, fining him 100 dollars. As a result of the Scopes trial, anti-evolution laws spread to other states in the South, even as public opinion in the USA turned against creationism. It is hard to believe that such a process could have occurred less than a hundred years ago! What would John Scopes' prosecutors have thought of the work of the physicists going on in Copenhagen and Göttingen?

July 1925, Göttingen

Walter Maurice Elsasser had just turned 21 when his boss Max Born gave him Clinton Davisson's curves. Unlike Davisson, Elsasser knew de Broglie's beautiful doctoral thesis and was absolutely fascinated by the idea that electrons could possibly also be waves. He now looked at the experimental results of Davisson and Kunsman with completely different eyes. No, these aren't scattering patterns of particles, as the two of them claimed. In this data, you can already see that it is the diffraction phenomena of a wave. Elsasser almost ran around the library shelves as he hastily picked out the papers with the well-known diffraction patterns of X-rays on crystal lattices and compared the curves printed there with Kunsman's measurement curves. Hell, what a resemblance! That is impossible. What a striking resemblance!

And here the student Walter Maurice Elsasser understood, was in fact the first in the world to understand, that the electron had already revealed itself as a matter wave much further back, namely in the work of Kunsman and Davisson. Kunsman's data was not the result of the scattering of particles but of the diffraction of a wave. What a beautiful discovery! He discussed this with Max Born. He spoke to James Franck. He wrote to Albert Einstein that the matter wave could be detected by diffraction of electron waves on metal surfaces. The regular rows of atoms on the surface of metals acted like a diffraction grating on the electron wave. And when Einstein wrote back: "Young man, you are sitting on a gold mine!" he wrote his first specialist article as a 21-year-old student: a preliminary communication in the magazine *Naturwissenschaft*, which appeared in the summer of 1925 under the title: "Remarks on the quantum mechanics of free electrons." It says:

> *If the lattice constant used is that of the platinum crystal lattice and, due to the relatively low penetration depth of the electrons, a plane lattice is used as a first approximation, this crude calculation yields values for the wavelength that correspond to the order of magnitude calculated according to de Broglie (to about 100%).*[28]

---

[28] Elsasser (1925).

Well, a match of around 100% is not really a proof. But at least it agrees with de Broglie's beautiful hypothesis in a preliminary way. Indeed, very preliminary. But nonetheless, such a clever idea that the likes of James Franck, Einstein, and the benevolent editors of *Naturwissenschaft* benevolently allowed it to slip through. This was the birth of the realization that waves of electrons with their wavelengths fit well with the lattice spacing in metals and should show diffraction phenomena in their rows of atoms. Thus it could be expected to find evidence of the matter wave here. De Broglie had been right!

August 1925, Cambridge

The mathematical physicist Ralph Fowler from Cambridge receives the proofs of Heisenberg's Heligoland article in mid-August. He had seen Heisenberg in July and asked for the paper to be sent to him. Heisenberg had given Max Born his article for review on July 10th and then immediately got on the train to Leiden and shortly afterward arrived in Cambridge, following an invitation from Ralph Fowler. On July 28th, Heisenberg gave his first lecture in Cambridge at the Kapitza Club. It was probably all a bit makeshift, a tiny college room filled with attendees, some of whom were sitting on the floor. His lecture had the strange title "Term-Zoology and Zeeman Botany." Fowler's PhD student, Paul Dirac, was supposed to attend but didn't. Heisenberg and Dirac did not meet at that time. Heisenberg hadn't actually wanted to talk about his Heligoland ideas, but Ralph Fowler had somehow coaxed it out of him and then made him promise when he left that he would send him the paper as soon as possible, and Heisenberg kept his word.

So now Fowler has the proofs of Heisenberg's article in his hand. Oh God, he must have thought, it's in German: "*Über eine quantentheoretische Umdeutung kinematischer und mechanischer Beziehungen*." Working through that would be excruciating. It's torturous even if you know German! And when he thinks of the word "excruciating," he immediately thinks of Paul Dirac. Nothing is too complicated for him. And so he puts the proofs in an envelope and sends them to Dirac at his home address, 9 Julius Road in Bristol, where he is staying for his summer holidays. In the top right corner of the first page, Fowler writes the words: "What do you think? I look forward to your opinion."

And now Dirac holds a paper in his hand that will not only change his life but the trajectory of all of modern physics. The paper that will finally, finally give him the truly original idea he has been looking for for so long. So what does Dirac do when he finally holds this piece of paper in his hands? He reads through it quickly, yawns, and puts it away. The whole thing is far too artificial. He shakes his head. Far-fetched. A theory without electron orbits! Nonsense! To focus only on what is observable. Useless! Dirac is completely unimpressed.

Dirac had returned to Bristol for the summer holidays after receiving a scholarship for another 3 years from the Royal Commission for the Exhibition. The Royal Commission, which still exists today and awards academic scholarships, was founded in 1851 by Prince Albert, husband and royal consort of Queen Victoria, to use the profits of the famous Great Exhibition of 1851 to "increase the means of industrial education and expand the influence of science and art on productive industry."[29] Dirac's application was supported by many eminent figures at Cambridge

---

[29] The Royal Commission for the Exhibition of 1851. (n.d.). Retrieved from https://royalcommission1851.org/

as well as the influential economist John Maynard Keynes. And so Dirac was initially well financed, fed, and happy. And Heisenberg's paper did not disturb his comfort.

But Dirac is not just a thinker, but someone who just can't let go of a thought, even after he's finished thinking about it. Who still thinks when everyone else has stopped thinking! The following anecdote is an example of this unique form of Diracian reflection:

> *Once, when some physicists were drinking tea together, Wolfgang Pauli put so many pieces of sugar in his cup that the colleagues sitting around him made fun of it. But not Dirac. He remains quiet and serious. His colleagues ask him about his position on the sugar problem. Dirac thinks about it and says: "I think a lump of sugar is enough for Professor Pauli." The subject changes. But after two minutes, Dirac says: "I think a lump of sugar is enough for everyone." The conversation continues. Dirac ends his train of thought and remarks: "I think the sugar lumps are made so that one is enough."*[30]

Everyone has already moved on to other things, but Paul Dirac sticks to his thoughts and only finishes them when he is completely sure that these thoughts are complete. So he reflects once more and naturally returns to Heisenberg's Heligoland paper, just as he returned to the question of the sugar lumps.

What fascinated him? The Heisenberg matrices had a strange property. There were matrices for location and matrices for momentum. And such matrices can be multiplied with each other, just as numbers are multiplied with each other. But with numbers, the order is irrelevant: three times four is the same as four times three. The factors can be swapped, so multiplication is commutative. But this was not the case with Heisenberg's matrices. The position

---

[30] von Weiszäcker (1984).

multiplied by the momentum was strangely not identical to the momentum multiplied by the position. Matrix multiplication was not commutative. The order of multiplication was essential. And this result of Heisenberg's work magically attracted Dirac! All of a sudden, he thought this was the key to a fundamentally new way of thinking about quantum physics. His biographer reports:

> *Several years later, his mother said in an interview that Dirac was so excited that he broke the rule of not telling his parents about his work and did his best to explain "noncommutativity" to them. He never tried again.*[31]

Dirac had worked on projective geometry for many years and therefore knew non-commutative quantities, i.e., quantities where the order of the quantities makes a difference during multiplication. He was thrilled to rediscover it here in a physical context, but at first he couldn't recognize its significance for the new quantum theory and didn't really know how to build on Heisenberg's ideas. For Heisenberg, the non-commutativity of his matrices was a rather disturbing discovery, but for Dirac it was the gateway to a new world. He suddenly realized that he had finally found a really original idea for himself!

September 1925, Göttingen

September was almost coming to an end when Max Born was finally satisfied. His and Pascual Jordan's article was actually finished. What he liked best was the short title "*On Quantum Mechanics.*" It sounds modest and yet as if something very fundamental is being negotiated here. The work,

---

[31] Farmelo (2016, p. 86).

which reached the *Zeitschrift für Physik* on September 27th, was, Born later said, the highlight of his research. Immediately after intensively devouring Heisenberg's Heligoland work at the end of July, he and Pascual Jordan had begun working on this paper. In the foreword, Jordan and Born initially heave a lot of praise on Heisenberg's work just to continue:

> *But from a formal, mathematical point of view, as he himself emphasizes, his considerations are only in the early stages. He only explained his hypotheses using simple examples and did not advance to a general theory. Favored by the fact that we were able to get to know his ideas in statu nascendi, we tried, upon completion of his investigations, to clarify the mathematical-formal content of his approaches and present some of our results here. They show that it is actually possible, on the basis given by Heisenberg, to build the structure of a closed mathematical theory of quantum mechanics in a strangely close analogy to classical mechanics, but while maintaining the features characteristic of quantum phenomena.*[32]

So that is the essence of this work: a closed mathematical theory of quantum mechanics. Written without haste, the ideas develop very slowly, didactically, and extremely artfully, almost like a textbook. Born and Jordan's first insight is that Heisenberg unknowingly talked about matrices in his article and that there are already many ready-made results in mathematics that are very useful here. Let's read about it in the Born-Jordan paper:

> *The mathematical basis of Heisenberg's analysis is the multiplication law of quantum theoretical quantities, which he developed through an ingenious correspondence analysis. The elaboration of his formalism, presented here, is based on the remark*

---

[32] Born and Jordan, Zur Quantenmechanik (1925).

> *that this rule is nothing other than the law of multiplication of matrices, well known to mathematicians. The square scheme, which is infinite on two sides, is the so-called matrix and is the representative of a physical quantity that is given as a function of time in classical theory. The mathematical method of the new quantum mechanics is therefore characterized by the use of matrix analysis instead of the usual number analysis.*[33]

Thus, the term "matrix mechanics" for the Heisenberg form of quantum mechanics is born. The second major result of this work can be found when you visit Max Born's grave in the city cemetery of Göttingen. Emblazoned on the tombstone in golden letters is a beautiful formula: the commutation relation of quantum mechanics. The product of the matrices of position and momentum is not exchangeable. Momentum times position minus position times momentum are the unity matrix times Planck's constant times the complex number i. This commutation relation which almost at the same time captivated the still entirely unknown student Paul Dirac, was already laid out in Heisenberg's paper, where it played the decisive role as the actual quantum condition but was not particularly clearly expressed. Heisenberg writes in his memoirs:

> *Born and Jordan now delved into the mathematical consequences of my work, this time without my presence, as I had been invited by Ehrenfest and Fowler to give lectures in Holland and in Cambridge, England. In just a few days, Born and Jordan discovered the crucial relationship pq - qp = h i/2 Pi, with the help of which the entire mathematical scheme could be made transparent.*[34]

---

[33] Ibid., p. 859.
[34] Hermann, Heisenberg (1976, p. 33).

Perhaps it should also be mentioned that Max Born spent the time when this important paper was developing in a sanatorium in Silvaplana, namely from mid-August to mid-September. During this time, Pascual Jordan kept him in suspense with an exhilarating exchange of letters. Jordan was later asked whether he actually wrote the paper alone, and replied:

> *It may indeed have been as you suspect—namely, that when Born and I met again in Göttingen, I had already pretty much completed the first draft.*[35]

You have to imagine this: This paper, for which Born is particularly famous, appears to have been largely written by Jordan, who was only 22 years old at the time. This was to be a common thread throughout Jordan's career: He was always overshadowed by others: by Max Born, by Paul Dirac, and mostly by Heisenberg. It is even suspected that his doctoral thesis was the actual inspiration for Heisenberg's uncertainty principle. His contributions were either overlooked or tacitly usurped by others without due credit. "Here rests someone who always stood in the shadow." That probably would have been the appropriate epitaph for Pascual Jordan.

Even when Born was back in Göttingen, Jordan continued to write letters, this time not to Silvaplana, but to Heisenberg in Copenhagen. Heisenberg had been back in Copenhagen since September 11th and was of course dying of curiosity to find out what Born and Jordan had worked out. Jordan briefly and succinctly summarized the results of the work in a letter dated September 12th. Heisenberg replied the very next day, immediately adjusting to the new situation and began to correspond eagerly with Jordan.

---

[35] Rechenberg (2010, p. 370).

Henceforth, letters flew back and forth between Copenhagen and Göttingen. Jordan, Born, and Heisenberg immediately continued with their great flight of ideas and, letter by letter, laid the foundations for what was later called the "three-man work," the third and final paper in the great series introducing matrix mechanics.

While work on the three-man paper began with Jordan's letter of September 12th, Born and Jordan concurrently were putting the finishing touches to their manuscript. On September 27th, Max Born not only sent the work to the *Journal of Physics*, but also sent a copy to Heisenberg in Copenhagen. Heisenberg immediately shows the work to his boss Niels Bohr:

> *Here, I received a paper from Born that I don't understand at all. It's full of matrices and I barely know what they are.*[36]

Which was definitely cheating a bit, because at that point he had already been working on the Born-Jordan work for almost 14 days and should have known the matrix formalism inside out. But it does not matter. At some point he surely internalized it, even had to master it, because to this day the name Heisenberg is intimately associated with the term matrix mechanics, so you better understand something about matrix algebra!

October 1925, Göttingen

In October 1925, work on the new quantum theory proceeded with great dedication. And in two places at the same time, but with neither knowing about the other. It seemed like Amundsen and Scott's race to the South Pole. Lonely

---

[36] Greenspan (2006, p. 135).

and alone the student Paul Dirac sat in Cambridge, working on his big idea. And in Göttingen, Messrs. Born, Heisenberg, and Jordan worked together and arrived at an all-encompassing theoretical draft that was to go far beyond the Born-Jordan and the Heligoland paper.

In September, while still in Copenhagen, Heisenberg had worked out a first-order perturbation theory and immediately sent it to Jordan. This perturbation theory became the actual starting point for the three-man work. Now began an intensive correspondence between Göttingen and Copenhagen. In their letters, the three gentlemen dealt with transformations from one coordinate system to another and wondered what that would do to their matrices. And then they figured out which transformation was particularly significant. One can guess: it is a transformation that leaves the beautiful commutation relation on Born's gravestone untouched. And because it behaves so beautifully, they called it a *canonical* transformation. Finally, in their letters they agreed why this canonical transformation was so important to them: you can now move on to coordinates that bring the energy matrix, i.e., the Hamilton matrix, into a diagonal shape and then the values of the energy levels of the quantum system being treated can be directly read off from these diagonal elements. So this is where the idea of eigenvalues first came into play. And Born had learned directly from David Hilbert how to deal with eigenvalue problems. So if you want to remember the central ideas of the three-man work, think of the keywords canonical transformation, eigenvalue problem, and diagonalization of the Hamilton matrix.

This is where a path that began in Heligoland has come to an end for the time being. In Heligoland, Heisenberg separated himself from the electron orbit as the actual descriptive quantity and instead introduced abstract matrices for the position and momentum of the electron that had to do with observable transition amplitudes. Then Born and Jordan found the position–momentum commutation relation as the essential basic equation of quantum mechanics. And in the three-man work, it becomes clear that you have to diagonalize the energy matrix, i.e., you have to deal with an eigenvalue problem in order to get to the discrete energy levels. So, instead of classically determining energies based on electron orbits, you now apply the newly found rule of quantum theory and determine energy levels as eigenvalues of a matrix. Born writes to Bohr:

> *However, I am aware that I lack the physical intuition that you (Bohr) or Heisenberg have and that all I can do to further the cause is to classify the physical regularities into a mathematical scheme. That's why I'm so happy now that Heisenberg's approaches fit beautifully with my obsession and that quantum mechanics can be given the form of an eigenvalue problem; In this way, it seems to me, one can quite easily overview all theoretical possibilities. I am very curious to see how far the principles formulated so far reach; it is probably possible that new principles will have to be added. But I am firmly convinced that the skeleton of the formulation is sensible.*[37]

And now? We are in the middle of one of the most important works in modern physics and what is Max Born doing now? Crazy as it may seem, he drives away! With the work half-finished, Max Born sets off on a long journey and leaves his highly talented assistants alone back in Göttingen. Wow! That is gutsy. Max and Hedi Born leave Göttingen on October 26th, leaving their children in the care of their grandfather and embarking on the ocean liner Westphalia on October 29th to travel to America for five long months.

---

[37] Letter from Max Born to Niels Bohr dated October 10, 1925 (Rosenfeld, 1977, p. 312).

At least Born had the "strange feeling" that he would miss a lot, and admitted in a letter to a friend that "it was simply the wrong moment to leave." He was undoubtedly correct about that.

October 1925, Cambridge

Oh, how beautiful Cambridge is! Especially in October, when the leaves on the trees turn so pretty. Dirac liked to take long walks around Cambridge on Sundays. He needed to. Ostensibly to forget about his work, but in fact to come up with new ideas. And so it happened that on one of these walks, he finally came up with the redeeming idea of how he could build on Heisenberg's Heligoland work, namely the idea that he could somehow draw on the mathematical concept of the Poisson bracket and then incorporate it into Heisenberg's quantum theory. He later said about this time:

> *I never went to theaters. I spent most of my time by myself, sitting working things out or going for walks. I used to spend every Sunday going for a long walk, a whole day walk, taking lunch with me…I found these occasions most profitable for new ideas coming. It was on one of those occasions that the possibility of ab-ba corresponding to a Poisson bracket occurred—on one of those Sunday walks.*[38]

Poisson brackets, named after Siméon Denis Poisson, a nineteenth-century mathematician, were a forgotten and somewhat esoteric formalism from classical mechanics: you can express the classical equations of motion very elegantly using these Poisson brackets. It actually had the same form as the position–momentum commutation relation in

---

[38] Dirac P. (1962).

Heisenberg's Heligoland work, which Born and Jordan also found so fascinating. That was Paul Dirac's sudden realization on his autumn walk: the great idea that these Poisson brackets could somehow be connected to the commutation relations of the Heisenberg matrices and that it should be possible to base quantum mechanics on them.

> *The idea came to me in a flash at first, I suppose, and of course caused a certain amount of excitement, and then of course came the reaction, no, that's probably wrong. It really was a very unsettling situation, and it became imperative for me to refresh my knowledge of the Poisson bracket.*[39]

His great misfortune, however, was to come up with such an idea on a Sunday, because all the libraries in Cambridge were closed. How infinitely painful it must have been for the young Dirac to come up with such a great idea and not be able to immediately read what exactly these Poisson brackets actually are. Twenty hours of waiting for poor Paul until the libraries finally opened again the following Monday morning. But every wait comes to an end and he immediately finds the book, which helps him check whether his idea is any good.

Even today, one can only read Paul Dirac's works with admiration. With him everything is always so clear and straightforward. Every line is important, one thing always flows neatly from the other, it is always fundamental and always fundamentally important, it is always abstract and always beautiful. Dirac's great achievement in 1925 was that he saw the position–momentum commutation relation (the formula on Max Born's tombstone) just as Born and Jordan did. That he was able to connect them in general with the Poisson brackets and in this way he was able to

---

[39] Farmelo (2016, p. 87).

bring all the equations based on the Poisson brackets from classical mechanics into quantum mechanics. Dirac now understands Heisenberg to mean that the variables that describe a dynamic system do not obey the commutative law of multiplication, but instead fulfill certain quantum conditions. One can then construct a theory without knowing anything about the dynamic variables other than the algebraic laws that govern them and can show that they can be represented by matrices. By linking Heisenberg's commutation relation with the Poisson brackets, he had, so to speak, built a transport bridge on which all the findings of classical mechanics could be transported into the new quantum mechanics. Thus, in effect a quantum theory based on commutation relations and Poisson brackets. This theory was far more ambitious and fundamental than Heisenberg's prototype. While Heisenberg had treated an electron in one dimension, Dirac's theory described the behavior of *all* quantum particles in *all* situations across *all* time. Just as Dirac himself was: absolutely principled. Let's listen to his biographer Farmelo:

> *His dream was that all the mathematics that Hamilton and others had used to transform Newton's theory of mechanics would have an exact counterpart in the new theory. If Dirac was right, physicists could use the infrastructure of "classical mechanics"—the stuff of hundreds of textbooks—in building the new theory that Heisenberg's boss Max Born had called "quantum mechanics" a year earlier.*[40]

Dirac was of course right and submitted his famous paper to the Proceedings of the Royal Society on November 7th.

---

[40] Farmelo (2016, p. 90).

November 1925, New York

On November 11th, the ship's siren on the Westphalia blared at five in the morning. They had finally arrived in New York. 14 days of travel to cover the 6200 km from Hamburg to New York. A surprisingly long time considering the blue ribbon for the fastest Atlantic crossing in 1925 was associated with an average speed of 26.06 knots, which for the Hamburg-New York route meant a journey time of 128 hours, or about 5 days. In any case, on November 11th, the Westphalia anchored outside New York Harbor and immigration officials began counting passengers and filling out their many forms. The city lay under a thick blanket of fog, with skyscrapers towering above, an almost ghostly sight.

At 10:30 am the steamer finally docked and Max and Hedi Born walked down the gangway. Their first day in America had begun! And when they finally got to their room at the Hotel Astor, they could hardly believe their luck. America! Imagine, Hedi, finally America! Henry Goldman, their benefactor, had sent his secretary who presently was tripping over himself trying to make their stay pleasant. And what a hotel it was! The bed linen was changed every day, there were carafes of ice water at meals, hundreds, even thousands of bellhops on duty who buzzed around and looked after you, entire boutiques integrated into the hotel lobby, and if you had a wish in your room, any wish no matter how unusual, you simply picked up a telephone! Yes, it's really true, there was a telephone with which you could then call reception directly. And now the happy couple began three intoxicatingly wonderful days in New York. On the first day, Hedi still thought she had ended up in a madhouse. But by the third day, she was deeply and genuinely distressed that they had to travel on to Boston.

Born had promised to give a total of 30 one-hour lectures at MIT in Boston, at least 20 of which were on the new quantum theory. And since he was so full of quantum theory, he spoke less about all the old stuff, the anomalous Zeeman effect or Bohr's atomic model, but rather over and over again about Heisenberg's Heligoland work, because he was fascinated by it and it was the only thing on his mind. Born originally wanted to stay for 5 months, including three months at MIT in Boston. But things turned out completely differently. His travel schedule soon filled up because quantum mechanics was suddenly in great demand and all the major physics centers between Boston and California suddenly wanted the German professor from Göttingen to speak. And so he went to General Electric in Schenectady and then to a total of 12 universities, including the universities in Ithaca, Buffalo, Chicago, Pasadena, Berkeley, Madison, Princeton, and Washington. Max and Hedi Born traveled almost 10,000 km across America. And just as the Pied Piper once lured children out of the city of Hamelin, Max Born drew physicists out of their studies and into his lectures. Thousands of professors, lecturers, and students in America, who had Max Born explain quantum mechanics to them firsthand. Around 1000 people attended his last MIT lecture on January 22, 1926. A young Harvard professor writes about Born:

> *It is a happy experience for American physicists that Professor Born was obliged to give this lecture on the dynamics of atoms just as the first presentation of the new matrix mechanics was appearing in Germany.*[41]

The Americans found the new quantum mechanics very, very exciting. And they really liked the fact that Max Born,

---

[41] Edwin Kemble quoted in Greenspan (2006, p. 141).

the co-inventor, personally taught and explained this new theory to them. Born quickly became famous and many students in his audience suddenly thought to themselves: this is the future, I want to be a part of it! For example, 23-year-old Carl Henry Eckart, the only son of German immigrants from St. Louis, heard Max Born in Pasadena and immediately and alone began working on the new theory over the winter of 25/26. Others thought: Should I perhaps do my PhD with Born in Göttingen? Should I go to the "Born of Quantum Wisdom"? Thanks to Max Born, the University of Göttingen suddenly became a true mecca of modern physics, at least in the USA. Just as we revere MIT today, people at MIT back then revered Göttingen. This was also the case with Linus Pauling, who had just received his doctorate, and suddenly thought completely differently about Germany and physics there and actually spent the year 1926 with Sommerfeld, Bohr, and Schrödinger. Interestingly, not only Born become famous in these months, but also Werner Heisenberg, because Born.

was scrupulous about giving Heisenberg the credit for the discovery of quantum mechanics.[42]

Perhaps he had exaggerated a bit too much: for years only Heisenberg was invariably cited in the USA, while the arguably more important work of Jordan and Born was only very rarely credited.

The trip to America not only made Born, Heisenberg, Göttingen, and the new quantum mechanics known and famous. No, it was also financially worthwhile for Max Born: four offers from the USA were the late reward for this lecture tour and ordeal. He was called by MIT, Cornell University, the University of Wisconsin, and finally Gerald

---

[42] Born M., Mein Leben (1978, p. 141).

Swope, the president of General Electric, waving a breathtakingly large check.

November 1925, Göttingen

Meanwhile, Heisenberg sat in Göttingen and had no idea that Max Born was making him a celebrity in faraway America. He sat and sat and sweated and tried to get to grips with the increasingly voluminous three-man work. Jordan remembers.

> *that the work, in terms of the material it was supposed to bring, was essentially finished before Born's departure, but was only finally formulated and compiled after his departure. At the time, Heisenberg was in charge of the final compilation.*[43]

Not an easy task for poor Heisenberg: firstly, because this type of physics was completely new for all three authors, but secondly because the three authors differed dramatically from one another in their nature and rarely agreed. Heisenberg puts it this way:

> *I still have serious doubts as to whether the problem of writing this three-man paper can even be solved in a finite amount of time.*[44]

But finally, on November 16th, it was really done. The three-man work is sent from Göttingen to the *Zeitschrift für Physik*. Heisenberg immediately wrote to Wolfgang Pauli:

---

[43] Letter from P. Jordan to B.L. van der Waerden from October 8, 1964. See van der Waerden (2013, p. 56).

[44] Letter from Heisenberg to Pauli dated October 23, 1925, see Klein and Toomer (1979, p. 251).

I tried my best to make the work more physical than it was and I'm half happy with it. But I'm still pretty unhappy about the whole theory and was so glad that you were so on my side with your view on math and physics. Here I am in an environment that thinks and feels exactly the opposite and I don't know if I'm just too stupid to understand math. Göttingen falls into two camps, one that, like Hilbert (or Weyl in a letter to Jordan), talks about the great success achieved by introducing matrix calculus into physics, the other, as James Franck says, that you will never be able to understand the matrices. I am always angry when I hear the theory referred to only under the name of matrix physics and for a while I seriously intended to delete the word matrix from the work entirely and replace it with another one, e.g. quantum theoretical quantity. (By the way, matrix is probably one of the dumbest mathematical words in existence.)[45]

You can hear Heisenberg's pride. He knows that he has really achieved something. But the reaction from the physics world was initially very cautious: the three papers on matrix mechanics were not particularly enthusiastically received. They are not very clear and seem far too mathematical to most people. Somehow no one understands them and no one can really see the benefits. Wolfgang Pauli gets to the point:

It must be possible to free Heisenberg's mechanics a little more from the torrent of formal scholarship in Göttingen and to expose its physical core even better.[46]

Well, what matters more now? When the masses applaud or when the great polymath Hermann Weyl confesses in a letter to Max Born: "Your approach to quantum theory

---

[45] Letter from Heisenberg to Pauli dated November 16, 1925, see Klein and Toomer (1979, p. 255).

[46] Letter from Pauli to Ralph Kronig dated October 9, 1925, see Klein and Toomer (1979, p. 247).

made a huge impression on me..." Sometimes it's simply a matter of waiting!

November 1925, Chicago

In November 1925 the music was not only playing in Göttingen. But also in Chicago. It was there that Louis Armstrong founded the "Hot Five" combo and recorded his first record with them. His voice is as unmistakable as his million-dollar smile. Whether with his trumpet or his singing voice, Louis Armstrong changed the face of music and culture so much that it is impossible to imagine the world without him. Up to this point, jazz mostly follows the "New Orleans style" of big band arrangements. Louis Amstrong takes the next step into music history with his band: the performance of solo improvisations! In doing so, he created the template for jazz improvisations that endure to this day, a cultural innovation that later played a leading role in the Harlem Renaissance.

In the 1920s, the world suddenly became interested in African-American art and culture. Armstrong and his virtuoso skills had a great influence on many leaders of this movement, to which he gave voice as a singer. Louis Armstrong's success was explosive. And success begets even more success. He was quickly offered supporting movie roles and was an almost universally popular public figure worldwide. The first real crossover artist, one would say today. Due to his immense popularity with both white and black audiences, Louis Armstrong became the epitome of the jazz age of the Roaring 20s. Along with Prohibition, jazz was an exciting and subversive part of the youth subcultures that flourished in the speakeasies of New York and Chicago in the 1920s. With his virtuosic musicianship and exuberant personality, Louis Armstrong played a crucial role in expanding the cultural reputation of jazz, helping to ease racial tensions and enhance the reputation of African-American art and culture. And this was at a time in history when racism thoroughly permeated American society. Louis Armstrong: a true giant of music history.

November 1925, Leiden

Two students from Paul Ehrenfest's group in Leiden, namely Samuel Goudsmit and George Uhlenbeck, propose that the electron has its own rotation, a spin. George Uhlenbeck was born in the same year as Wolfgang Pauli, 1900, in Jakarta, where his father had served in the Dutch East India Company. Sam Goudsmit, still called Samuel Goudschmidt at the time, on the other hand, was born in 1902, just like Heisenberg. He was 23 years old and hailed from The Hague. He later had a remarkable career as a physicist in the USA and rose to prominence particularly toward the end of the Second World War as part of the so-called Operation Epsilon, an operation by the Allies to clarify whether or not the Germans had had the ability to build an atomic bomb. As part of this operation, he captured Werner Heisenberg as an American officer and then brought him to Farm Hall in England, the same Heisenberg with whom he had a friendly correspondence with while still a student.

Uhlenbeck and Goudsmit discovered the spin of the electron in November 1925. Of course, that is technically incorrect because there was nothing left to discover. Pauli had long ago assigned a half-integer spin quantum number to the electron, but without labeling it "spin" and without a clear interpretation of this fourth degree of freedom. Uhlenbeck and Goudsmit only put the descriptive interpretation on top: that the electron supposedly rotates around its own axis, hence the word "spin," an abbreviation for "intrinsic angular momentum." The paper appeared in the German magazine "*Die Naturwissenschaft*" in November 1925. In 1971, Sam Goudsmit gives a lecture and tells the story himself:

And that was it: the spin; thus, is was discovered, in that manner. Of course, we told Ehrenfest about it and then summer was over and I went again to Amsterdam and various episodes followed. Naturally, I found it wonderful, because in the formalism which I knew it fitted perfectly. And the rigorous physics behind it I did not fathom. But Uhlenbeck, being a good physicist, started to think about it……. "A charge that rotates"……? He claims that he then went to Lorentz and that Lorentz replied: "Yes, that is very difficult because it causes the self energy of the electron to be wrong". We had just written a short article in German and given to Ehrenfest, who wanted to send it to "Naturwissenschaften". Now it is being told that Uhlenbeck got frightened, went to Ehrenfest and said: "Don't send it off, because it probably is wrong; it is impossible, one cannot have an electron that rotates at such high speed and has the right momentum". And Ehrenfest replied: "It is too late, I have sent it off already". But I do not remember the event, I never had the idea that is was wrong because I did not know enough. The one thing I remember is that Ehrenfest said to me: "Well, that is a nice idea, though it may be wrong. But you don't yet have a reputation, so you have nothing to lose". That is the only thing I remember.

Well the [paper] was submitted and published. Directly, the next day, I received a letter from Heisenberg and he refers to our "mutige Note" (courageous note). I did not even know we needed courage to publish that. I wasn't courageous at all. I think I still have Heisenberg's letter. In it he writes a formula ……… I did not understand a bit of it. And then he says somewhere: "What have you done with the factor 2?" Which factor? Not the slightest notion, and the formula given without derivation.[47]

So whether the idea that the electron actually rotates makes sense has to be very much questioned in light of Lorentz's comment. Furthermore, it would also have a speed at its equator that would be many times the speed of light. Impossible! But it's just an image in your head that you may have, but don't have to have. To each their own. Factor two, which Heisenberg pointed out, is more problematic. The effect of the new magnetic spin quantum number on level splitting is exactly twice as large as the effect when using the magnetic orbital angular momentum quantum number. This factor 2 is actually a relativistic effect, which Goudsmit and Uhlenbeck of course could not explain and which was only discovered later.

Wolfgang Pauli himself had already brushed off a corresponding idea from the physicist Ralph Kronig in January 1925, a fact that held against him subsequently. As bad as that may sound, Pauli's skepticism was entirely appropriate and understandable. In the end, the spin is simply a purely quantum mechanical property of the electron. From a quantum mechanical point of view, everything looks like a classic angular momentum, but it has absolutely nothing to do with the rotational movement of a ball of finite size around its own axis. How the spin ultimately comes about remains inexplicable in classical physics, no matter how much we still long for a clear interpretation. An explanation for the spin can actually only be obtained from the Dirac equation and it is based on relativistic quantum mechanics.

Contemplating all of this, one wonders what, then, Gouldsmit and Uhlenbeck actually found. An image in

---

[47] "The discovery of the electron spin," S.A. Goudsmit, https://www.lorentz.leidenuniv.nl/history/spin/goudsmit.html

your head that you can easily envision, but which unfortunately is wrong. In any case, Wolfgang Pauli never received the recognition he deserved for his part in the discovery of spin. It was only much later that Goudsmit himself understood why Pauli always greeted him with the cryptic remark that he could afford not to be cited. But somehow Pauli wasn't truly angry with the two of them. You can see this when you read the CV written by Pauli for his professorship in Zurich. It says:

At the end of 1924 I wrote a paper which, among other things, contained a general theorem relating to atomic structure, which has proven to be very fruitful in unraveling complicated spectra and has since been widely quoted in the literature. This work continued in the direction I had begun to pursue in Copenhagen; It also brought various suggestions to the Dutch physicists Goudsmit and Uhlenbeck, who brought this problem to a provisional conclusion through their theoretical discovery of the magnetic properties of free electrons.[48]

So: Pauli discovered the spin at the beginning of 1925, but did not call it that, but described it as a "peculiar, classically indescribable type of ambiguity" and added it to the theory as the fourth degree of freedom of an electron. Ralph Kronig then wanted to interpret this as the self-rotation of an electron, which Pauli talked him out of, but which was then made up for by the students Goudsmit and Uhlenbeck, who also became famous for it, even though they themselves actually knew that this interpretation was no good at all. Very confusing!

---

[48] Enz & von Meyenn (1988, p. 42).

November 1925, Zurich

Another year had passed since Louis de Broglie received his doctorate. The main idea of his work: electrons behave like waves. Matter waves. The miraculous idea spread rapidly throughout the world of science because none other than Albert Einstein promoted it by mentioning it in his papers and explicitly praising it. And so the idea found its way into pretty much every theoretical seminar in the world and, of course, also to Professor Schrödinger at ETH Zurich, who gave a theoretical seminar on de Broglie's dissertation there on November 23, 1925.

Peter Debye was of course also present on November 23rd as he never wanted to miss a lecture by his friend. As Schrödinger was eagerly pontificating about the wave nature of the electron, Debye made a remark that was rather casual and quite dry. One of those remarks that everyone fears because it is so obvious a question, yet the speaker had never asked it himself. Debye's remark stopped Schrödinger in his tracks: Debye mentioned that he had already learned as a student of Sommerfeld's that you needed a wave equation in order to really be able to talk about waves. And where, pray tell, is the wave equation for de Broglie's waves?

And at that moment the good Lord must have held his breath in anticipation to see whether Schrödinger would recognize the great opportunity of his life. He could have just casually shrugged his shoulders. But he didn't. He immediately saw how important his friend's question was, dropped everything and went to work on this question immediately. It would be the question of his life, and just 3 weeks later, before Christmas, he invited people to another seminar, where Peter Debye was again sitting in the front row. He stood confidently in front of the audience and slowly and solemnly spoke the words: "My colleague Debye suggested that one should have a wave equation; Well, I found one." Unfortunately, the wave equation was not yet correct. For this, he would have to go on a skiing vacation to Arosa.

December 1925, Arosa

There in Arosa he sat at the turn of the year and pondered and tinkered and pondered some more and finally built his equation. Crafting or building—you can't call it anything else. The Schrödinger equation is a differential eq. A differential equation is a very special equation. You can solve a normal equation for the unknown and then get a number. But if you don't just want to relate numbers to one another, but rather whole functions, then you invoke a differential equation. The solution to a differential equation is not a number, but rather a whole function. Schrödinger now wants to deduce a solution in reverse, from known functions to the differential equation that is still unknown to him and the world, i.e., ultimately to construct it. To do this, he takes the electromagnetic prototype of a wave equation, which he specifically modifies so that the law of conservation of energy of a non-relativistic electron, the Planck relationship between energy and frequency, and the de Broglie relationship between momentum and wavelength can be accommodated in this equation. In the end, he has a wave equation that has de Broglie's matter waves as solutions.

Incidentally, he was not alone in Arosa, but with an "old girlfriend" from Vienna, while his wife stayed at home. Nobody knows who this friend was. Schrödinger's biographer explains very laboriously which of Schrödinger's many friends it certainly wasn't, only to then state:

> *Whoever may have been his inspiration, the increase in Erwin's powers was dramatic, and he began a twelve-month period of sustained creative activity that is without a parallel in the history of science.*[49]

You can easily imagine that the girlfriend had little joy during this vacation with the withdrawn and brooding Erwin. "Come on, Erwin, let's hit the slopes! I could just as well have stayed in Vienna!"

Most of what he will publish in the next 6 months he completed in Arosa all by himself; only a few smaller mathematical details are still missing, for which he needs the advice of his Zurich colleague Hermann Weyl. This equation by Erwin Schrödinger is now the central equation of all quantum mechanics and can be found in every book on quantum mechanics. But also in the arcade courtyard of the University of Vienna, where it is emblazoned on the plinth below the marble bust of Erwin Schrödinger.

December 1925, Frankfurt

It's not just quantum physicists who are making big leaps. In the world of chemistry, too, discoveries are being made that will change the course of world history, and as with quantum mechanics, Germany is at the center. On December 2, 1925, the five largest chemical companies in Germany merged to form a mega-corporation: IG Farben. IG Farben would exist only 27 years, but at least enough time for the world to see it as a driver of technological progress. The guiding spirit of this idea was Carl Duisberg, who had long advocated for the merger and is now the first

---

[49] Moore (1992, p. 195).

chairman of the supervisory board, the so-called "Council of Gods."

With the outbreak of World War II, IG Farben became a tool of the Nazi regime. Almost half of the company's hundreds of thousands of workers were slave laborers from Auschwitz and other camps. Among other things, they developed and produced Zyklon B, the chemical used in the Nazi gas chambers, and took part in human medical experiments on camp inmates. Without the technical expertise and production capacity of IG Farben, Germany would not have been able to go to war in 1939. This story reveals the sometimes dark implications of scientific discoveries: that they cannot always be separated from how they get used. In fact, many of the discoveries in the history of quantum physics found their first practical application in the construction of the atomic bomb, a weapon of unspeakable horror.

December 1925, Boston

Max Born sits in his room at MIT in Boston and goes through his mail. He still has to prepare the lecture for the afternoon and has no time to spare. He looks at one of the letters. A preprint in the "Proceedings of the Royal Society" from Cambridge. By a Paul Dirac. From whom? Paul Dirac? I don't know, never heard of him. He reads the editor's note:

Paul Dirac is the 1851 Exhibition Senior Research Student of St. John's College, Cambridge and submitted his paper on 7. November 1925. The paper was communicated by Ralph Fowler.[50]

---

[50] Dirac P. A. (1925).

A Cambridge student? Max Born is completely amazed, he starts to read and almost falls off his chair! The word "perplexed" only roughly describes Max Born's reaction. He and his boys in Göttingen have just completed a piece of work that can hardly be surpassed in terms of genius, when he receives a paper from a single student from Cambridge, completely unknown to him, who not only has understood everything far better but actually sums it up much more elegantly. It was completely inexplicable to Born how a Cambridge student could have done this alone. But he is even more amazed and astonished that this completely unknown student even got wind of the matter. He has no idea how this student got Heisenberg's paper. Born writes in his memoirs:

> *Dirac's work was one of the biggest surprises of my life [...], the author seemed to be very young, but everything was perfect and admirable in its own way.*[51]

Dirac gave his work a very confident title: "The Fundamental Equations of Quantum Mechanics." After reading the article, it immediately becomes clear that he wasn't exaggerating. Even if many physicists thought that his findings were too far removed from the real world, Paul Dirac himself was quite relaxed about this. For him, it was only a matter of time before his equations would be confirmed by real experiments. Of course, at first glance they seemed quite unconventional. But Dirac simply knew that his equations correctly described the quantum behavior of particles under all possible circumstances.

His ideas differed from those of Heisenberg in one important point: While Heisenberg thought that one should be able to start with Bohr's early formulations of the

---

[51] Born M., *My Life: Recollections of a Nobel Laureate* (2015, p. 226).

quantum world and somehow build on them, Dirac dared to make a radical new start. He made it unmistakably clear: quantum mechanics is an independent mechanic for the quantum world, completely equal to classical mechanics, to which it shows a strict correspondence. And it was precisely this correspondence, this parallelism to classical mechanics that proved to be groundbreaking for the young field of quantum mechanics, since most scientists at the time were far more familiar with the world of classical mechanics than with the rather foreign-seeming ideas coming out of Copenhagen and Göttingen.

At home in Cambridge, Dirac's ideas made little impression. Most of his colleagues there had no interest at all in the new quantum physics. For them, Dirac was a brilliant but dry mathematician. The new quantum mechanics still had to show its value in the experimental world, because that was the real test for any new theory. But perhaps this was Cambridge's problem rather than Dirac's, because at Cambridge theoretical physics, mathematics, and experimental physics were strictly separated from each other. The intellectual culture of continental Europe, however, was very different in that these three areas were much more intertwined.

It is no wonder then that Dirac became instantly known through this paper. Everyone interested in quantum theory read his work and reacted just like Max Born. Since this Dirac had been an utterly unknown entity, people in Göttingen, Copenhagen, and Munich were completely astonished that such work suddenly appeared out of nowhere. One could sense that in this very moment, a consequential mathematician and physicist first saw the light of day.

Heisenberg, too, was surprised. On November 23rd, a few days after receiving the copy of Dirac's proofs, Heisenberg responded with a long letter to Dirac, who, despite his competitive feelings, would become a good friend. He wrote:

> *I read your extraordinarily beautiful work on quantum mechanics with the greatest interest and there can be no doubt that all your results are correct, provided one believes in the new theory at all. Hopefully you are not saddened, however, by the fact that some of your results had already been found here some time ago.*[52]

Heisenberg was referring to the Born-Jordan paper, which Paul Dirac hadn't even noticed. It must have been a brutal disappointment for Dirac. He thought he had got his hands on the coattails of fame, but it turns out that the Göttingen team was faster. Heisenberg writes another letter to Dirac:

> *Please do not take these questions of mine as criticism of your wonderful work. I have to write an article about the current state of the theory and am still amazed at the mathematical simplicity with which you have conquered this problem.*[53]

Heisenberg appears quite fair and comradely. Seems that he is extending his hand to Dirac and wants to integrate him into his group. This presumably was important for Dirac as he was pretty much alone in England, while Heisenberg had an incredibly well-established network: Bohr, Kramers, Pauli, Born, and Jordan, plus all the mathematical support that was at his fingertips everywhere in Göttingen. All of Göttingen here and over there just the lone fighter Dirac. While Heisenberg was still reserved in his spontaneous letters to Dirac, later on as an old man he emphasized the importance of Dirac's work in an interview:

---

[52] Farmelo (2016, pp. 92–93).
[53] Ibid., p.93.

So I should say it was a combination of Born and Jordan's paper on the one hand, and Dirac's paper on the other hand, which really made the whole thing.[54]

In any case, through Dirac, in addition to Göttingen and Copenhagen, Cambridge also came into play as the place where quantum mechanics originated.

While just 7 years prior they would have been forced to shoot at each other in the trenches, an international group of very young scientists worked together across borders, cultures, and languages on an incredibly challenging theory that no individual could have created on their own. This theory indeed was to become a genuine international collaborative effort. And this brief timeframe in the 1920s was ideal for such an undertaking: impossible before 1918 and even more so after 1933. Peace had finally spread across Europe. As possibly the most beautiful fruit of this peace, quantum mechanics exudes the so-called "spirit of Locarno." The groundbreaking treaties signed in October 1925 in the little Swiss town of Locarno confirmed the borders of France, Germany, and Belgium, established a non-aggression pact between the signatory nations, and finally enabled Germany to join the League of Nations. In return, the Allied troops withdrew from the Ruhrgebiet, an area of great economic importance for Germany. Locarno—a great success of negotiations, for which the three lead negotiators all received the Nobel Peace Prize. And yet, Locarno also laid the foundation for later sprouting of dissatisfaction. Poland, for example, also would have liked a non-aggression pact with Germany, which would have benefited them in 1939. But the fatal mistake was the choice of guarantor: namely Mussolini's Italy. The Italians had their own imperial ambitions, and when their invasion of Ethiopia in 1935

---

[54] Heisenberg W., Interview of Werner Heisenberg (1963).

failed to elicit a response from the League of Nations, Hitler saw his chance to undermine Locarno by remilitarizing the Rhineland. The spirit of Locarno was fine while it lasted, but Europe still had to wait a good two decades for a truly lasting peace.

1926, Antithesis: Wave Mechanics

January 1926, Hamburg

The fruit of the peace of Locarno, the new quantum mechanics, is at this point still not ripe. Hard, sour, and completely inedible. A partial birth. Everything is far too abstract. The Göttingen theorists around Werner Heisenberg have to hear that, and Paul Dirac also hears it in Cambridge. So what needs to be done to convince the critics? It's very simple: With your theory, you calculate the most famous and experimentally most accurately measured spectrum in the quantum world: the line spectrum of the hydrogen atom, described by the so-called Balmer formula, which reflects the distances between these lines in the spectrum. Anyone who can conjure up the Balmer formula from their formalism will make their theory truly credible and should immediately win over a number of skeptics. That's what Paul Dirac thinks in Cambridge, that's what Wolfgang Pauli thinks in Hamburg. And both of them go to work individually and cross the finish line almost at the same time.

On January 22nd, 1926, Paul Dirac delivered his essay "Quantum Mechanics and a Preliminary Investigation of the Hydrogen Atom," but on January 17th Wolfgang Pauli submitted his article: "On the hydrogen spectrum from the standpoint of the new quantum mechanics." Again Dirac had come in second.

By successfully applying Heisenberg's matrix mechanics to the hydrogen problem, Wolfgang Pauli has found the most important piece in this part of the puzzle, a real partial success for the new theory. Only now are most physicists convinced of the new matrix mechanics. The formula already played an important role in Niels Bohr's by now outdated model. He is one of the first to congratulate Pauli:

> *Dear Pauli!*
> *I was very pleased to hear from Kramers that you had succeeded in deriving the Balmer formula. I am very much looking forward to hearing more about this act and hope that, as you promised Kramers, you will write to me about it soon.*[1]

Bohr wrote that on November 13th. And Heisenberg's congratulatory letter doesn't take long either:

> *I don't need to tell you how excited I am about the new theory of hydrogen and how much I admire that you came up with this theory so quickly.*[2]

Even Max Born in distant America reported on Wolfgang Pauli's success in his 17th lecture in Boston. And news of victory from Hamburg has already reached Ralph Fowler in Cambridge. He writes to Niels Bohr on December 4th:

---

[1] Letter from Bohr to Pauli dated November 13, 1925, see Klein and Toomer (1979, p. 254).

[2] Letter from Heisenberg to Pauli dated November 3, 1925, see Klein and Toomer (1979, p. 252).

I have of course been intensely interested in Heisenberg's new theory, thanks to Dirac I have expected rapid development on those lines, but Pauli's work on hydrogen has got ahead magnificently—much faster than I expected to hear of at first.[3]

Even though he only submitted his work in January, Pauli had actually already finished his work at the end of October, otherwise, he would hardly have been able to accept his congratulations in November. It is also interesting to note the role that letters obviously played in spreading new ideas back then.

Concealed behind all the appreciative words is a certain relief that Wolfgang Pauli is now laying hands on the new theory. Indeed, so far he has only been on the sidelines with respect to Heisenberg's great Heligoland idea. Conversely, Heisenberg always included him in many, many letters and asked him for advice and comments at every step. Pauli was always well informed and closely integrated into the Göttingen group as a critical interlocutor.

But he just didn't participate. He somehow stayed on the sidelines from where he gossiped critically about the "Göttingen erudition." Gossip, but not pitching in! That's simply annoying! At some point, Heisenberg got really tired of Pauli's snide remarks and at the beginning of October he wrote to Pauli without holding back:

I have to give you another sermon, and I apologize if I continue in Bavarian: it really is a pigsty that you can't stop ranting. Your eternal complaints about Copenhagen and Göttingen are simply a screaming scandal. You'll have to let us be, since we're not maliciously trying to ruin physics; If you accuse us of being such big donkeys that we can never achieve anything physically new, that may be true. But then you are just as big a donkey, because

---
[3] Letter from Fowler to Bohr dated December 4, 1925, see Klein and Toomer (1979, p. 253).

you can't do it either .... (The dots mean a curse lasting two minutes!) No offense and best regards W. Heisenberg.[4]

In short it means: you gossip on the sidelines and you can't do it yourself! That must have left an impression on Pauli because in any case, a few days after this letter he tackled the hydrogen problem in just a few days. Seems like that was the tone necessary to get someone like Wolfgang Pauli fired up. Not an easy person to deal with!

January 1926, Soho, London

But now the story needs a short break. And so make yourself a cup of coffee, put your feet on the table and pick up the newspaper once again. This time it is the "Durlacher Tageblatt," or, to appreciate the full title in its baroque length, the "*Durlacher Tageblatt: Local newspaper for the city and the former Durlach district; Pfinztäler messenger for Grötzingen, Berghausen, Söllingen, Wöschbach and Kleinsteinbach.*" The article with the headline: "Television is here!" jumps at you.

> *The television that scientists and researchers around the world have been dreaming of for half a century has now become a practical reality. One British inventor, John Logie Baird, was responsible for bringing the device to commercialization. After years and years of patient research, he saw his work crowned with success when, in January 1926, forty critical men of science, members of the Royal Institute, witnessed the first demonstration of a television in his small laboratory and were*

---

[4] Letter from Heisenberg to Pauli dated October 12, 1925, see Klein and Toomer (1979, p. 250).

astonished to see living human images transmitted from one room to another.[5]

The first transmission of moving pictures. It is January 26, 1926. John Logie Baird surely was relieved and the gentlemen present in his laboratory in London's Soho district surely were impressed. The "televisor" actually works. Logie Baird, an engineer from Glasgow. His earliest experiments consisted of knitting needles, an old chest, bicycle lights, and an old hatbox. He took his invention around the whole of Glasgow. Everyone showed him the door. They thought he was crazy! Eventually, he came across a thallium sulfide cell developed in the USA to produce grayscale images. And now finally, on January 26, 1926, he had reached the first, but probably most critical stage: the images of his two ventriloquist dummies "James" and "Stooky Bill" were transmitted. Back to the Durlacher Tageblatt:

After these demonstrations, Baird spent two more years developing and perfecting his invention. During this time he managed to broadcast television through telephone lines between London and Glasgow and in wireless fashion between London and Neuyork [New York].

How beautiful, this word "Neuyork," with which the Durlacher Tageblatt probably wants to ensure that the significance of this news can be understood everywhere, especially in the towns of "Grötzingen, Berghausen, Söllingen, Wöschbach and Kleinsteinbach." The following explanation serves the same purpose:

Television will therefore actually travel over distances of many miles and over all the obstacles in between, and events that take

---
[5] Durlacher Tageblatt from October 4, 1928, see https://www.deutsche-digitale-bibliothek.de/

place in the distance will be seen at the same moment, i.e. enjoyed from afar, just as distant sounds can be heard by means of wireless telephony.

Moving images from all over the world can now be transmitted into people's living rooms using radio waves. Sitting in an armchair, you look at flickering images in a small box and are presented with images of events that originate in entirely different places. What an immense broadening of horizons! Quantum mechanics and the telescreen were born almost at the same time. Also of interest.

But the break isn't over yet, the coffee mug isn't empty yet. Next, you reach for the *Vorwärts* from November 1925:

Solving the robbery-murder in Stralsund.

The perpetrators were arrested and confessed to three crimes.

As we reported at the time, on July 13th of that year in Stralsund, an elderly Miss Kaiser was found murdered and robbed in the apartment of the postal assistant Fiedler on Seestraße. Fiedler and his wife were away and had given their apartment to Miss Kaiser, a relative, to look after. In particular, five albums with valuable stamps were stolen, a collection that was very well known in philatelic circles in Stralsund. It turned out that the murdered woman had been lying dead in the apartment for a long time. The robbers had disappeared without a trace. Chief Criminal Inspector Gennat's staff has now managed to fully investigate the bloody crime and put the two perpetrators behind bars.[6]

In Berlin during the Roaring Twenties, crime had reached levels that were previously unimaginable. And so, in January 1926, the first central homicide department was founded under the leadership of Ernst August Ferdinand Gennat, the "fat man from Alexanderplatz." Gennat's new

---

[6] Vorwärts from 11/11/1925, page 3.

department revolutionized the work of the criminal police, developed new methods for securing evidence and crime scenes, consistently used all the possibilities of modern forensics and soon had sensational success with a clearance rate of almost 95%. Nicknamed "Full Earnest," Ernest Gennat also developed what we today call "suspect profiling" and coined the term "serial killer." Gennat's work in Berlin made him an international star. Celebrities such as Charlie Chaplin and Thomas Mann visited his laboratory. Police forces from all over the world came to learn about his new methods. He was already a legend during his lifetime, was portrayed in films and became a role model for all future television commissioners. With Gennat, the crime thriller also enters the world stage. Logie Baird is responsible for the technology of the new age with the television, Ernst Gennat for the programming!

The mug is empty, the newspaper is finished. Back to work! Back to quantum mechanics!

January 1926, Zurich

Erwin Schrödinger published his Arosa wave equation in four articles in the journal *Annalen der Physik*. All four are referred to as a "communication" and all four have the same heading: "Quantization as an eigenvalue problem." The first communication reached the editorial team on January 27, 1926. One would expect that Schrödinger would begin his crowning work of the century with a justification and general discussion of his wave equation. But far from it. Instead, he goes straight into the hydrogen problem, i.e., into the quantum mechanical treatment of the problem of a single electron in the force field of a single proton, just as if the initial equation, the Schrödinger equation, had long been well known and not worth mentioning. The derivation of his equation remains somehow unclear, a fact that he himself points out again later. Schrödinger doesn't want to talk about his equation, but rather about the properties of its solutions. And this is what he writes right at the beginning:

> *In this communication, I would first like to use the simplest case of the non-relativistic and undisturbed hydrogen atom to show that the usual quantisation rule can be replaced by another requirement in which there is no longer any mention of "integers". Rather, the integer results in the same natural way as the integer of the number of nodes of a vibrating string. The new view is generalisable and, I believe, touches very deeply the true nature of the quantum rules.[7]*

What fascinated Schrödinger about the whole story was indeed the manner in which the quantum numbers emerged from this calculation and explains the title of the four communications "Quantization as an eigenvalue problem." In prior years, the quantum numbers always somehow dropped out of the sky; now they materialized free and easy. The Schrödinger equation describes de Broglie's matter wave for certain potentials of the electron. Here in this first communication from Schrödinger, the potential was that of a positive nucleus, a Coulomb potential in which the electron was somehow spatially trapped. While previously it was believed that electrons orbited around the nucleus, Schrödinger now let the electron be a matter wave that formed as a standing three-dimensional wave around the nucleus that held it. The electrons in the atom can only form very specific, fixed wave patterns, just like a clamped string, and precisely because it is clamped, can only realize certain

---

[7] Schrödinger, Quantisierung als Eigenwertproblem (Erste Mitteilung) (1926).

vibration patterns. This ultimately depends on the boundary conditions. In other words: The boundary conditions now result in a fixed sequence of these standing waves, the eigenfunctions, which are differentiated from one another by integers, the quantum numbers, which were ultimately the result of the boundary conditions of the eigenvalue problem. The eigenvalues for the energy eigenfunctions are then exactly the energy values that also resulted from the Bohr-Sommerfeld model.

February 1926, Munich

Schrödinger sends his treatise to the editor of the *Annalen der Physik*, Wilhelm Wien, in Munich and writes:

I would be very grateful if you would be so kind as to give the work to Privy Councillor Sommerfeld for his perusal before it goes to press, after it has received the 'Received' mark. Professor Sommerfeld has always shown such a kind interest in my most insignificant works that I am delighted to be able to present him with one that will really interest him.[8]

Sommerfeld replies to Schrödinger on February 3rd:

What you write, in your essay and letter, is terribly interesting. I was in the process of creating an outline for lectures in London that was in the same tone. Then, like a thunderbolt, your manuscript arrived.[9]

---

[8] Letter from Erwin Schrödinger to Wilhelm Wien, January 26, 1926.
[9] A. Sommerfeld in a letter to Erwin Schrödinger dated February 3, 1926.

On the Same Day, the Excited Sommerfeld Wrote in a Letter to Pauli

A manuscript by Schrödinger for the Annals arrived here. Schrödinger seems to find exactly the same results as Heisenberg and you, but in a completely different, completely crazy way, not matrix algebra, but boundary value problems. Surely something sensible and definitive will soon emerge from all of this in some form.[10]

Pauli Later Commented on This in a Letter to Jordan

I believe that this work is one of the most important things that has been written recently. Read it carefully and reverently.[11]

Meanwhile, Schrödinger was already writing his next communication, which reached the editor of *Annalen der Physik* on February 23, 1926. It reads like a modern physics book and raises the question of how so many fundamental topics can be gathered in so few lines. He particularly emphasizes the analogy between the wave theory of light, which passes into radiation optics in the limit of small wavelengths, and his wave theory of matter, which passes into classical mechanics in the limit of small de Broglie wavelengths. No wonder that his theory ever since has been called wave mechanics.

At this point I do not want to ignore the fact that Heisenberg, Born, Jordan and some other outstanding researchers are

---

[10] Letter from Sommerfeld to Pauli, February 3, 1926, letter 118a, see Klein and Toomer (1979).

[11] Letter from Pauli to P. Jordan, April 12 1926, see Klein and Toomer (1979, p. 315).

currently attempting to eliminate the quantum difficulty, which has already achieved such remarkable successes that it is difficult to doubt that there is at least some truth to it. Overall, Heisenberg's experiment is extremely close to the present one, as we have already mentioned above. The method is so completely different that I have not yet been able to find the link. I harbor the definite hope that these two initiatives will complement each other, one helping where the other fails.[12]

At some point between the second and third communication, Schrödinger still finds the time to describe the connection between Heisenberg's theory and his own. Heisenberg's matrix mechanics and his wave mechanics, Schrödinger opines, are equivalent theories: two different descriptions of quantum mechanics but of equal value. The paper[13] reaches the *Annalen der Physik* on March 18th.

However, this does not result in the blessing of Heisenberg, who at least initially rejects Schrödinger's work. He later confesses to Pauli that he thinks it is "crap." An antipathy that, by the way, is entirely mutual. In one of his papers, Schrödinger writes that he.

feels put off, if not outright repulsed, by the algebra and the lack of clarity of Heisenberg's theory.[14]

Wow! The word "repulsed" in a scientific article. This is unheard of! One can see that the wave mechanics under Schrödinger and the matrix mechanics under Born and Heisenberg are by no means pulling in the same direction. Quite the opposite: in the spring of 1926, a tough battle

---

[12] Schrödinger, Quantisierung als Eigenwertproblem (Zweite Mitteilung) (1926).
[13] Schrödinger, Über das Verhältnis der Heisenberg-Born-Jordanschen Quantenmechanik zu der meinen (1926).
[14] Ibid.

between the parties begins over who actually has the better grasp of the conditions in the atom.

February 1926, Cambridge

And now Paul Dirac was getting going. The man had a way of thinking that, according to Rudolf Peierls, "reaches its goal via a straight line where others start to swerve." In addition to the article on the hydrogen problem, he wrote two other papers in the first 4 months of 1926 which dealt with Heisenberg's new theory. Dirac was trained as an engineer, a background that he was able to combine with his mathematical skills in a very unique way. Doing abstract mathematics while being able to live with approximations and always keeping an eye on the physical application: that was Paul Dirac. His biographer Famelo puts it this way:

> *Dirac's method was so confusing because his science was an unusual mix—part theoretical physics, part pure mathematics, part engineering. Dirac had the physicist's passion for learning the fundamental laws of nature, the mathematician's love of abstraction for its own sake, and the engineer's insistence that theories must produce useful results.*[15]

Dirac himself says about the importance of his engineering training for his view of mathematics:

> *I think that this engineering education has influenced me very much in making me learn to tolerate approximations…..I rather got to the idea that everything in nature was only approximate, and that one had to be satisfied with approximations, and that science would develop through getting continually more and more accurate approximations but would never*

---

[15] Farmelo (2016, p. 96).

attain complete exactness. I got that point of view through my engineering training, which I think has influenced me very much.[16]

This approach to approximations constituted a conflict with Heisenberg and the other young scientists who were working on these problems and were much more guided by experimental results. Dirac, on the other hand, always worked out the theory first, next applied approximations, and only then checked whether he could make meaningful predictions about quantum behavior.

At the beginning of 1926, the Göttingen folks, i.e., Heisenberg, Born, and Jordan, were generally recognized as the pioneering group of the emerging quantum theory. But Dirac's work also began to make waves. The reproduction of the Balmer formula was only a first, small, partial victory for the young revolutionaries. The lack of further experimental results that could have confirmed the theory remained the main problem, for the Göttingen team as well as for Dirac. Unlike the Göttingen people, Dirac thought from the very start that this was solely because Einstein's special theory of relativity had not yet been integrated into quantum theory. And so he began working on that in the spring of 1926.

These were turbulent times in Cambridge, by the way. In the spring there was a general strike that lasted 9 days and paralyzed economic activity across the country. The strike even reached the remote halls of Cambridge, where many students volunteered as strikebreakers, took jobs as tram drivers or dock workers—so many that the university had to postpone final exams. Ultimately, almost 1.7 million workers went on strike, the last general strike in the history of the UK. Paul Dirac didn't care at all. He spent his time

---

[16] Dirac P. (1962).

writing up his doctoral thesis. When he was almost finished, he came across Schrödinger's early work. Should he address it? Somehow integrate it into his doctoral thesis? Or just ignore it? Ignore it, Dirac decided and put a full stop after the last sentence of his work.

February 1926, Berlin

A quick note reminding the reader that our serious physicists were also people of flesh and blood. Surely all of them had seen Charlie Chaplin's latest film and laughed their heads off like the rest of the audience: The Gold Rush, one of the truly great cinema classics. The *Dance of the Rolls* is legendary: a fork in each of two rolls, which Chaplin then makes dance on a table. In the summer of 1925 in American cinemas, in December 1925 in London, but only in February 1926 in German cinemas. The *Berliner Zeitung* reads: "The final act produces the most painful laughter that has ever emanated from a cinema. It's heartbreaking. And yet it's overwhelmingly funny!" And the *Filmkurier* reads: "The most beautiful film in the world, the absolutely, absolutely most beautiful film ever made." At the premiere in Berlin, the unimaginable happens: during the performance, the wildly clapping audience forces an encore of the *Dance of the Rolls*!

A completely different kind of film was also released in Germany during these months: *Battleship Potemkin*. Made in the Soviet Union, because the 1920s were a culturally rich time there as well. An avant-garde movement flourished, which was only halted by the brutal suppression that took place under Stalin. Artists such as Kandinsky and Malevich explored abstraction in painting, and writers such as Andrei Bely and Vladimir Nabokov explored new ideas such as Freudian psychoanalysis in their novels. And in cinema, Sergei Eisenstein achieved global success with *Battleship Potemkin*. The film is set during the Russian Revolution of 1905 and is about the mutiny of the crew of the battleship Potemkin. The mutineers head to the port of Odessa, where their anger sparks an uprising against the tsar. The whole thing culminates in a famous scene: The Odessa Steps, where Tsarist soldiers brutally attack the insurgents as they march down the 200 steps of the famous Odessa Staircase. In many countries, including the USA and Great Britain, the film was banned for years, censored, or edited repeatedly due to its strong emotional impact. While Eisenstein gives tragedy a new stage, comedy is Chaplin's thing. You can choose to laugh or cry till you feel purged, clear your head, leave your own grief behind and suddenly love your time again.

April 1926, Göttingen

Everyone, yes, everyone had shown up. James Franck, Pascual Jordan, Werner Heisenberg, Friedrich Hund, Carl Runge, Max Delbrück, Walter Elsasser, John von Neumann, Richard Courant, David Hilbert, and on and on and on. Everything with rank and name in Göttingen had come to Max Born's welcome party on Planck Street. So many illustrious guests all at once! Because after five long months in the USA, the friendly, the gentle, and the universally loved Max Born was back in Göttingen. Everyone wanted to see him, everyone wanted to hear about his impressions of America.

But Max Born only had eyes and ears for his children. He had really missed them dearly and had gone out of his way to bring back special gifts from the USA. But the errand had not been that easy as he didn't want to send out the hotel concierge. He wanted to give them something completely unknown in Germany, and thus also completely unknown to him. He couldn't have given the concierge a clear directive! And so Max Born himself, loaded with shopping bags, went through the big American cities to find presents for his children.

And just as all the world-famous physicists and mathematicians poured into the house, Born had just opened his big gift bag: 12-year-old Irene and 11-year-old Gritli were now wearing Indian headdresses, 5-year-old Gustav was wearing a—thank God!—perfectly fitting cowboy suit. And the children ran around enthusiastically and gratefully, which made Max Born incredibly happy. He could hardly take his eyes off them. Only now did he take the main gift, the state gift, out of his bag. An electric model train! The children were thrilled, and father and son immediately set up the new train in the middle of the large living room while the guests poured themselves a glass of punch and watched the two. Some of the distinguished guests must have secretly thought: "And that's why I come over? To watch father and son put together a train set?" But no one said anything, everyone was cheerful and relaxed.

And when the tracks were finally set up, the locomotive and wagons were placed on and the transformer was connected, Max Born asked for silence and imitated the sound of a drum roll increasing the tension, because: now the regulator on the transformer was turned up. And? Still drumroll! And? Oh no! Absolutely nothing happens. Nothing! The locomotive simply doesn't start. Gustav bursts into tears, the two girls look disappointed from under their headdresses. But thank God the crème de la crème of German physics is in the room! And so now all the jubilant physicists and potential Nobel Prize winners in their suits descend onto the floor and debate the question of why the wretched train can't get going. Is it the transformer? Are the wires loose? And while there is a heated and committed discussion, Hedi and little Irene are just quietly going through the pieces of track one by one until they have found and removed the defective one. And now the train is finally running! Thank heaven! Hedi then made a final remark that further increased the general embarrassment. The example showed that physics lacked women, she said. How could it be that only men developed quantum mechanics? If only we had a fair system where the other half of a country's talent reserve was also mobilized, where women could also lend a hand in the development of the new theory! How much more could then have been achieved! Where were all the female counterparts to the Heisenbergs and Paulis of the time? The group looked puzzled and embarrassed, the aspiring Nobel Prize winners turned away from the matter and returned to the punch. In those years, physics in Göttingen was primarily theoretical and less practical, and in any case dominated by men.

Only once the party was long over and the children and wife were already in bed was Max finally able to unpack his suitcase. He wanted to truly arrive at home and as quickly as possible find his way back to his beloved everyday life in Göttingen. And the best way to speed up this process is to quickly empty the suitcase and take it to the attic. At the bottom of the suitcase, he finds the manuscript of an article. For the editor of the *Zeitschrift für Physik*, Pascual Jordan had given it to him in October last year, shortly before his departure for the USA. Please, he should read it and, if he deems it to be good, send it on to the journal. For heaven's sake, Born thinks, I totally forgot! He hadn't penetrated these deep layers of the suitcase during the trip and had completely lost track of the article. Born hopes that the article does not contain anything of fundamental importance and that there is no harm in publishing it belatedly. He starts reading immediately. His hopes are in vain. The article contains something fundamentally important, something that every physicist knows today, namely the derivation of what is called Fermi-Dirac statistics, i.e., the statistical distribution of identical particles with half-integer spin, so-called fermions. And the name "Fermi-Dirac Statistics" gives you an idea of the drama. Enrico Fermi had already discovered the distribution and published it in an Italian newspaper in February 1926 and immediately afterward in the *Zeitschrift für Physik*. Born was deeply ashamed, not just that evening, but throughout the rest of his life. Later he said:

> *I hate Jordan's politics, but I can never undo what I did to him. When I returned to Germany six months later, I found the paper at the bottom of my suitcase.*[17]

Because of Born's scatterbrained nature, Jordan was denied the recognition he deserved. Pascual Jordan's paper containing this ever-so-significant result could well have been published in November 1925. In which case the distribution would nowadays be called the "Jordan distribution." Max Born looks at the floor, ashamed. How embarrassing! By the way, what Paul Dirac's name is doing in the "Fermi-Dirac distribution" will be explained shortly.

---

[17] Born M., Mein Leben (1978).

April 1926, Zurich

There is massive trouble in the Schrödinger house: Annie creates a scene that's really something. Does it have anything to do with that unknown lady on the skiing holiday in Arosa? Hadn't the rules of the game back in Zurich been utterly clear to everyone involved? A good, stable balance had been established as far as relationships between men and women were concerned.

There were four well-known scientists in Zurich. In addition to Schrödinger, these were the physicists Peter Debye and Paul Scherrer and the mathematician Hermann Weyl. With Peter Debye in particular Schrödinger maintained a warm friendship throughout his life. Hermann Weyl, the famous mathematician, physicist, and philosopher, who was considered one of the last mathematical universalists, was another important friend and colleague. These three scientists were certainly a crucial reason that Schrödinger accepted the appointment at ETH Zurich in 1921, surely in addition to the fact that none other than Albert Einstein had held his chair before him. "ETH Zurich"—That sounds impressive today, big and shiny. But this university started very small. It only received the right to award doctorates in 1908, just four years before Einstein accepted a professorship in theoretical physics there. And when Einstein earned his diploma as a "mathematics teacher" there in 1900, he was what we today call a vocational schoolteacher. But ETH was still the "Federal Polytechnic School" and what we would probably call a technical college today. Yes, nota bene, the scientific legend Einstein did not study at a university, but at a technical college.

But back to Schrödinger. When you look at the photos of these great scholars, Peter Debye, Paul Scherrer, Hermann Weyl, and Erwin Schrödinger, you are in awe of how intelligent and serious and important they appear. But when you read about their marriages and partnerships, they all seem to have discovered and practiced open relationships. And this is where the four women in question come into play. Annie, Schrödinger's wife, had a long-standing and open relationship with Hermann Weyl, while Weyl's wife Hella made no secret of her relationship with Paul Scherrer. Thirdly, Schrödinger had an extramarital affair with the wife of his colleague Arthus March and had a daughter with her. Unfortunately, it is completely unknown who Debye had any relationship with. And it is not known whether Scherrer's wife in turn had a relationship with Arthus March, which would have closed the circle.

In any case, people at ETH Zurich not only gave free rein to their thoughts but also to their bodily desires. Walter Moore, Schrödinger's biographer, puts it in his own way:

Their Zürich circle provided ample opportunities for amorous adventures, and if Erwin cared to venture outside the usual academic liaisons, both Weyl and Debye were available as competent guides to the uninhibited night life of the city.[18]

Hard to believe! Schrödinger was a very good-looking man who had many different love affairs with other women throughout his life: he ended up having three children with three women, none of them with his wife. But there is also a darker side to Schrödinger's life including misogyny and sexual abuse. He is known to have pursued several relationships with underage girls, leading some institutions today to remove his portrait and rename lecture halls that previously bore his name.

---

[18] Moore (1992, p. 191).

May 1926, Zurich

May 1926, the third communication from Erwin Schrödinger in his series "Quantization as an eigenvalue problem." In this 53-page work, Schrödinger develops the stationary perturbation theory and calculates the splitting in the Stark effect. He cannot yet address the anomalous Zeeman effect; to do this, he writes, one must incorporate the Uhlenbeck-Goudsmit idea, i.e., the spin of the electron, into wave mechanics. Wolfgang Pauli takes care of that a year later.

And while Schrödinger is on a roll, producing paper after paper within a very short space of time and thereby writing his way to world fame, his audience is already starting the big debate. Schrödinger doesn't even know what his wave actually means. With the greatest naivety, he believes that the electron is no longer a particle, but that it will completely dissolve and move through space as a wave in a completely smeared state. While that is an image easy to picture, the vast majority of physicists know that this actually cannot be the case in light of the preceding history. Worth mentioning as an example is Old Lorentz, who remarks on wave mechanics in a letter to Sommerfeld dated May 27th:

If we decide to completely dissolve the electron, so to speak, and replace it with a wave system, it has both a disadvantage and an advantage. The disadvantage, and a critical one, is this: what we assume of the electron of the hydrogen atom, we must also assume of all electrons in all atoms; we need to replace them all with wave systems. But how am I supposed to understand the phenomena of photo-electricity and the escape of electrons from heated metals? Here the particles appear quite nice and intact; How did they, once dissolved, come together again?[19]

---

[19] Letter from H. A. Lorentz to E. Sommerfeld, May 27, 1926, see Przibram (1963, p. 41).

May 1926, Copenhagen

It is May 1st and thus the day Dr. Werner Heisenberg commences his job as a lecturer at the Institute for Theoretical Physics at the University of Copenhagen. He is now the official assistant to the world-famous Niels Bohr, thus assuming the position of "His Eminence the Cardinal" to "Pope Bohr." His predecessor, Hendrik Anthony Kramers, accepted a professorship at Utrecht University at the beginning of the year. Heisenberg had just left Copenhagen in mid-October of the previous year, and by then there were no longer any leaves hanging on the trees. And now, on May 1st, the trees are still bare. Makes you wonder just how short the growing season is up here in the North. No wonder there are so few deciduous trees here. When will it be green? Everything is already blooming in Munich and in Naples, where he also had already been this year, the cherry trees blossomed in March.

But that's not the only thing that might surprise Heisenberg. He might be in constant wonder about his world, his life, the rapid staccato of events. The fast pace is without a doubt a sign of this strange time. Heisenberg had just read in the newspaper about Henry Ford and the introduction of a standardized 40-hour week. Ford originally wanted to introduce a 6-day week, but due to pressure from unions and fears of the spread of Soviet revolutions, they settled on a 5-day week. So this was the beginning of a world of assembly line work, where every worker is part of a much larger process. A revolution in manufacturing and the beginning of mass production and consumerism.

The high rhythm, the hammering staccato, the rapid succession of events, the many, many opportunities that present themselves: Werner Heisenberg cannot curtail his amazement. How does all this happen? What world does he inhabit? Born had barely secured him the coveted private lecturer scholarship at the Georgia Augusta University in Göttingen at the beginning of the year when he received the offer from Bohr for the position in Copenhagen. And as soon as he accepted, a letter from the University of Leipzig fluttered to his doorstep, offering him a professorship in theoretical physics—at the tender age of 24! How enticing. But Heisenberg wavers. Professor in Leipzig or assistant in Copenhagen? Which is better?

A few days before he took up his duties in Copenhagen, Heisenberg was in Berlin. He spoke in the legendary physics colloquium at the University of Berlin and the great scholars from Einstein to Planck, from Meitner to the Nernst listened to him. "The bigwigs were all there!" he reports to his parents. "Everyone unanimously advised me to go to Copenhagen!" Above all, it was Einstein who advised him, from whom he received the most honorable invitation to a discussion in his private apartment following the colloquium.

By the way, it is worth at this point to briefly mention that the beautiful Einstein word "quantum egg" causes confusion. Einstein is said to have used it once in a letter to Ehrenfest dated September 20, 1925. But then again Max Delbrück claims to have heard it when he entered the Berlin lecture hall where Werner Heisenberg was supposed to give a lecture. Albert Einstein and Nernst met there. And Nernst, according to Delbrück, asked what was next. And Einstein—Max Delbrück remembers this very clearly—supposedly said to Nernst:

Heisenberg has laid a quantum egg and now let's see if it can be hatched.[20]

---

[20] Fischer (2022, p. 152).

So what now? Was the word quantum egg used in September of '25 or April of '26? Or perhaps Einstein made use of the term on two occasions? Leaving aside this extremely disturbing inconsistency in the historical record, it should quickly be noted that the 20-year-old Max Delbrück decided to switch to physics because of Heisenberg's quantum egg lecture and then became a student of Max Born in Göttingen.

But back to Heisenberg, who was advised by Berlin celebrities to forgo the professorship in Leipzig and go to Niels Bohr instead. He writes to his parents:

Today I finally rejected the call to Leipzig, because one is not put into the world to hold offices, but to see whether one can achieve something in science.[21]

It was simply clear to him that quantum mechanics is not yet completed and that a lot of physics is still needed to really understand and interpret what has already been achieved. And he prefers Bohr's way of thinking to the mathematical discussions in Göttingen or even the intellectual solitude in Leipzig. Understandable! Yet a year later he went to Leipzig after all, but that is in the future. It's still May first and Heisenberg begins his stint in Copenhagen. He is supposed to help Bohr look after visiting scientists and visitors and also give a one-hour lecture per week on a wide variety of subjects of modern physics, albeit in Danish, which he apparently managed quite well.

The novelty of his new position in Copenhagen was a provided apartment. Inconceivable! But Niels Bohr's assistant received an official apartment. This will only make sense if you know the story of the Physics Institute in

---

[21] Heisenberg to his parents, Berlin, April 29, 1926 (Cassidy, 1995, p. 272; Hirsch, 2003, p. 102).

Copenhagen, which Niels Bohr himself laboriously and with a lot of patience and blood, sweat, and tears procured from the government, and which he then planned himself and in which he and his family also lived. A three-story building on Blegdamsvej, with the large letters UNIVERSITETS INSTITUT FOR TEORETISK FYSIK emblazoned above the entrance gate. To the right of the large entrance hall is the auditorium, as well as a library and a cafeteria. And, yes, also an official apartment for the assistant:

> *My official apartment at the institute is very nicely furnished; Bohr lent me a piano; The ceiling is not painted, as it was in Göttingen, and all the domestic horrors: Venus de Milo, etc. are missing… Every few days, Bohr and I take riding lessons together so as not to get bogged down in physics.*[22]

The riding lessons were absolutely necessary because Werner Heisenberg was indeed in danger of drowning in physics. Because he, too, wrote paper after paper, not surprising, given that so much suddenly becomes clear after such a fundamental breakthrough. In March, together with Jordan, he applied the new quantum mechanics to the problem of the anomalous Zeeman effect by incorporating the spin hypothesis of Uhlenbeck and Goudsmit into his theory. The whole thing was a rather peculiar construct: on the one hand, matrix mechanics with its farewell from any old notions and its commitment to complete abstraction. And on the other hand, spin, which everyone imagined as a rotating ball. Very, very peculiar! But the result: an impressively well-fitting theoretical explanation of the fine structure of the hydrogen lines. And now in Copenhagen Heisenberg wanted to turn his attention to explaining the

---

[22] Heisenberg to his parents on May 5, 1926 (Hirsch, 2003, p. 103).

spectra of the helium atom: now there are two electrons that make up the atom. Let's see if the child prodigy can achieve that too!

May 1926, Pasadena

Carl Henry Eckart, a newly graduated physicist at the California Institute of Technology, worked through the winter of 1925–1926. He had been completely alone. He heard Max Born at Caltech in Pasadena in December and immediately withdrew from the world and went into retreat. His idea came to him during Max Born's lecture. Namely: to represent quantum mechanics using its own operator formalism. When Schrödinger's first article appeared in the *Annalen der Physik* in February 1926, Eckart immediately recognized the overwhelming importance of this discovery and asked himself like everyone else: how can two theories as different as Heisenberg's and Schrödinger's say the same thing? One with infinitely large matrices, the other partial differential equations. How does this work? There must be a general mathematical framework from which both formalisms emerge as two equivalent representations.

Supervised by no one, without any opportunity for a clarifying conversation, far away from the stimulating European science centers, Carl Eckart is now working on this very problem in Pasadena and… finds the solution: the link between the proper functions of the Schrödinger equation and the matrices of the Jordan-Born-Matrix algebra. How excited and happy he must have been while writing his article, which reached the "Proceedings of the National Academy of Sciences" on May 31, 1926. And what a relief when the journal finally confirmed receipt of the manuscript. He had decided not to celebrate until the published document was in his hands. And when he happily, but still alone, went into town to drink cold German wheat beer at his favorite pub, he decided to quickly stop by the magazine room of the Caltech Physics Institute. Oh, if only he hadn't! There he found the latest edition of the *Annalen der Physik* and in it yet another communication from Erwin Schrödinger, dated March 18th. Schrödinger showed the equivalence of Heisenberg's presentation and his own, i.e., the result of Carl Eckart, which he had just planned to celebrate with a crystal-clear Hefeweizen from Bavaria. Eckart was deeply disappointed and only drank naturally cloudy beer that evening. But plenty of it.

May 1926, Cambridge

Dirac's dissertation has a title that is as simple as it is succinct: "Quantum Mechanics." It is the first dissertation in the world to be submitted on this topic. The defense is set for mid-May. Paul is 23 years old, and since he only got his hands on Heisenberg's paper in September of '25, pretty much everything in this work was probably written between September 1925 and April 1926. Dirac's work before September 1925 was not included at all. In his doctoral thesis, Paul Dirac wanted to present something definitive, honest, fundamental, and to the point. So, he concentrated exclusively on the essentials: and in his eyes that only pertained to the work of the last six months. As expected, the dissertation was a great success. The examiners in Cambridge were deeply impressed. On behalf of the faculty's examination committee, its chairman wrote a personal letter to Dirac by hand, congratulating him on the exceptional importance of the work.

June 1926, Cambridge

So, there stood the Vice-Chancellor of the University of Cambridge, quickly adjusting his ermine collar, which had seen better years and looked as if the ermine that belonged to the collar had been present at the laying of the foundation stone of St. John's College. But tradition and strict adherence to form, that was Cambridge. As true then as it is today. If someone wanted their doctorate from the University of Cambridge, they had to endure this quasi-religious ceremony, all in fine thread, groomed and spurred, complete with gown and bow tie. Needless to say, Paul wasn't particularly enthusiastic about the event. He almost drifted off when the Vice-Chancellor began to deliver a speech in Latin in a nasal tone, which hardly anyone in the room mastered well enough to be able to follow. But when Paul's gaze fell on his father Charles, who was sitting in the audience and looked at him sternly, he quickly decided against dozing off after all. Charles and Florence had made the journey from Bristol, which had cost a considerable amount of money as Charles had repeatedly noted. But at least the long-suffering Florence, who had never gotten over Felix's death, seemed to feel something akin to happiness again and looked with obvious pride at her unique, but now unfortunately also only, son Paul. She was worried that he wasn't taking care of himself and kept urging him to take a rest over the summer. A piece of advice that Dirac deliberately ignored, because if he was looking forward to the quiet of the summer holidays, not to take a vacation, but to finally get on with his work.

June 1926, Oxford

In the same month, a hundred kilometers southwest of Cambridge, in Oxford, Clinton Davisson is at a conference and for the first time really looks in depth at the de Broglie idea of matter waves. Max Born, the supervisor of Walter Maurice Elsasser, gives a lecture and explains the ideas of his students, also showing the diffraction data from the Davisson-Kunzman experiment. Davisson sits in the audience and is only slowly beginning to understand that the article by Walter Maurice Elsasser, which he studied last fall, is probably not as foolish as he initially thought. In an autobiographical note, he explicitly states that it was by no means Elsasser's paper that led him to his famous experimental proof of the matter wave. But that is difficult to imagine, given that Elsasser obviously had the crucial idea that the de Broglie matter wave could be detected by electron diffraction on a metal surface. And it was Elsasser who made it clear to Davisson how close he was to prove the matter wave with his existing experimental setup and that all he really needed was a "well-defined crystal object," i.e., a single crystal. How can he claim that Elsasser's article was meaningless to him? Let's leave that point open. In any case, back at Bell Labs in New Jersey, Davisson announced his "Program of thorough search" and was able to demonstrate a personal quality that indeed distinguished him. His remarkable attention to detail. Maybe the 21-year-old from Alsace was a bit smarter than he, but in terms of thoroughness, no one in the world could compete with him.

June 1926, Hamburg

Having returned from his post-doc in Copenhagen in the fall of '23 and previously been an assistant at the University of Hamburg, Wolfgang Pauli received an offer of an associate professorship in theoretical physics at the University of Leipzig in June 1926. Recall, the poor University of Leipzig had already unsuccessfully approached Werner Heisenberg with the same offer and had been blown off by him. To recruit Heisenberg before Pauli: what chutzpah! Did Wolfgang Pauli think favorably of that? The University of Hamburg, thank God, reacts promptly, awards him the title of professor, and gives him a special teaching assignment to entice the already famous Pauli to stay in Hamburg. Which he did, leaving Leipzig empty handed once again.

But who actually was Wolfgang Pauli? Max Born once said of Wolfgang Pauli that he was a "genius, comparable only to Einstein himself." And then he went one better and added: "In purely scientific terms, perhaps even greater than Einstein!" In light of such greatness, isn't it comforting to know that even geniuses have their weaknesses? Because as a human being, Pauli was an extremely problematic person. The literature is full of stories and anecdotes about Wolfgang Pauli that bring him back to life in his utter brutal directness. But probably the most unsparing characterization of him originates with Carl Gustav Jung, whom Wolfgang Pauli consulted as an analyst in the 1930s. Wolfgang Pauli's contradictory, rebellious, and difficult persona is summed up here:

> *His judgment appears cold, unbending, arbitrary and ruthless: as clear as the inner structure of his thoughts is to him, it is unclear to him where and how they fit into the real world; when he falls for people who don't understand him, he gathers evidence of the unfathomable stupidity of people; or he develops*

into a misanthropic bachelor with a childlike heart: he seems bristly, aloof and haughty: he is afraid of the female sex.[23]

Bristling, aloof, and haughty: that's Wolfgang Pauli. As a human being, Wolfgang Pauli was in danger of slowly going downhill in the years of his greatest successes, in the *Sturm und Drang* years of quantum mechanics. His nightlife, his visits to bars and his drinking had taken an extreme turn for the worse. Hamburg is known for its notorious and seedy nightlife strip of St Pauli. And Pauli may have taken the word "St Pauli" as a personal invitation to him. Tobias Hürter writes in his book "Das Zeitalter der Unschärfe":

He begins a double life: by day he is the honest university lecturer, by night he is the bon vivant. When it gets dark, he wanders through the bars and variety shows of St. Pauli, where the counters are sticky with spilled beer and the walls are dark with tobacco fumes, where Josephine Baker, who is banned from performing in Munich, dances her Charleston.[24]

One automatically thinks of the Robert Louis Stevenson characters *Dr. Jekyll and Mr. Hyde*. Furthermore, he is of the completely mistaken opinion that alcohol makes him a more interesting person. Pauli says:

…that drinking wine suits me very well. After the second bottle of wine or champagne, I tend to adopt the manners of a good companion (which I never have when sober) and can then, under certain circumstances, make a tremendous impression on those around me, especially if they are female![25]

---

[23] Hürter (2021, p. 288).
[24] Ibid. p. 284.
[25] Ibid.

In fact, he has a massive drinking problem, gets into fights, and becomes aggressive. But he doesn't see any problem at all. He describes his time in Hamburg as probably the happiest years of his life, perhaps because he had colleagues here, the physicist Otto Stern and the mathematician Erich Hecke, with whom he got along extremely well. In 1928, he accepted a professorship at the ETH Zurich, but his drinking problem didn't improve there either.

June 1926, Copenhagen

Werner Heisenberg has hay fever again. That's why he fled to Heligoland last year. Now he's sitting here in Copenhagen with Bohr and is thinking about escaping again. Heligoland is inaccessible from Copenhagen. How about Norway? High in the mountains without any pollen yet.

> *In the meantime, I would like to go to Norway because of the well-known hay fever and, and aside from mountain climbing, do a quantitative calculation of the helium spectrum. Why not drive a steam roller for a change? We'll probably see each other somewhere in Germany in the summer.*[26]

He wrote this in a letter to Wolfgang Pauli on June 8th. What does he mean by "driving a steamroller"? Probably that he intends to solve the riddle himself! As we know, helium has two electrons in its atom. Extending his theory of quantum mechanics to two electrons will not be easy. You have to flatten everything like a steamroller to get through it.

---

[26] Letter from Heisenberg to Pauli dated June 8, 1926, see Klein and Toomer (1979, p. 328).

The letter of June 8th is particularly famous because of its ending attacking Schrödinger. It's not just old Hendrik Lorentz who has serious doubts about Schrödinger's views. Heisenberg also takes issue with Schrödinger's interpretation that the electron is a charge distribution smeared across space in the form of a wave. In the letter dated June 8th he writes:

> *By the way, an unofficial note about physics: the more I think about the physical part of Schrödinger's theory, the more disgusting I find it. Imagine the rotating electron, whose charge is distributed over the entire space with the axis in a fourth and fifth dimension. What Schrödinger writes about the beauty of his theory... I think it's rubbish. The great achievement of Schrödinger's theory is the calculation of the matrix elements. The connection with de Broglie has not yet been discovered. But excuse the heresy and don't tell anyone. Greetings to all of Hamburg physics! W. Heisenberg.*[27]

June 1926, Zurich

Back in Zurich Schrödinger is still unaware of this opposition movement. On June 21, 1926, he submits his fourth and final paper in the wave mechanics series. Finally, he finds the time-dependent Schrödinger equation. Why did it take so long? Probably because he struggled with the fact that he could only show a complex wave function, underscored by a quote from the work:

> *There is no doubt that there is currently still a certain amount of difficulty in using a complex wave function. If it were fundamentally unavoidable and not just a simple calculation, it would mean that there are basically two wave functions that*

---

[27] Ibid.

only together provide information about the state of the system, a somewhat unpleasant conclusion.[28]

Reading all four missives in succession, you get the sense that Schrödinger only bit by bit understood what a fundamental discovery he was tracking. From communication to communication, the tone becomes more confident. What might he have thought upon realizing that he would be the one who would be privileged to find and announce one of humanity's greatest discoveries?

From de Broglie's doctoral thesis, published in 1924, there is a direct path to Schrödinger's four papers in the spring of 1926. Precisely the leaps of an idea from one mind to another to another are so interesting: De Broglie has a bold thought, Einstein recognizes and propagates it, Schrödinger lectures on it, Debye poses a question and Schrödinger then formulates an equation. An equation that is the basis for a vast new field of knowledge. An equation that made Schrödinger world famous.

Interestingly, Schrödinger's fellow students had already anticipated this rise to world fame. It happened in the winter semester of 1907–1908. Erwin Schrödinger is 20 years old and enrolled at the University of Vienna exactly one year ago to study physics and mathematics. A student of the same age, Hans Thirring, is sitting in the brightly lit library of the mathematical seminar when a blonde young man enters the room and the person sitting next to him at the table nudges him excitedly: "that's Schrödinger!" Hans Thirring writes:

I had never heard the name before, but the respect with which it was said and the facial expression of my peer made such an impression on me that from the very first encounter I became

---

[28] Schrödinger, Quantisierung als Eigenwertproblem (Vierte Mitteilung) (1926).

convinced, and over the course of years ever more so: That one is something special ... And long before he achieved his great success with the development of wave mechanics, it was completely clear to his inner circle of friends that one day something very significant could be expected from him. We clearly saw in him a fiery spirit at work that joined forces with the drive to explore that breaks through the constraints of narrow subject areas in order to independently and in new ways query nature.[29]

When Erwin Schrödinger passed his final exams with brilliant grades in the spring of 1906, it was a given for him that mathematics and physics should accompany him through life. He was 19 years old. The first 20 years of his life in his beloved hometown of Vienna were so carefree and easy. In this respect, his life could definitely be compared with the life of Prince de Broglie, who, like him, decided to study physics at the Sorbonne at the age of 19 years. Schrödinger, by the way, always considered him a theoretical bungler with a brilliant idea, which seems rather ungrateful in that he owes his world fame solely to de Broglie's idea.

June 1926, Göttingen

A free electron moves through space and collides with a single atom. A collision between electron and atom. A so-called collision process, as Born's colleague James Franck had investigated experimentally. How exactly can this collision process be described quantum mechanically? Max Born had already suffered this question in Boston. Wanted to address it with the very matrix mechanics he had co-developed, but simply could not pull it off. Big annoyance!

---

[29] Thirring (1947, p. 106).

The resulting equations simply could not, could not be solved. A tangled ball of wool had been created. Born had to give up. A very big annoyance.

But now there was the Schrödinger equation, a completely fresh equation that still showed the strains of its birth. As soon as he returned from the USA, the "Born of quantum wisdom" pounced on Erwin Schrödinger's papers and practically devoured them. And at first, he was once again profoundly annoyed. In the USA, he had written a complicated and generally little-noticed paper with Norbert Wiener, a young mathematician from MIT, in which linear operators instead of matrices now played the decisive role in the formulation of quantum mechanics. After reading Schrödinger's work, he immediately realized that he and Wiener had stopped one tiny step short of discovering wave mechanics. Just one small step and he would have had it! Big, big annoyance! Born later writes that wave mechanics.

> …is the best example of my fate, being on the verge of an important discovery and then letting it slip out of my hands.[30]

Well, Max thinks, wasn't meant to be. Get up and carry on! And when he finally gets over his anger, he quickly realizes that Schrödinger's equation is extremely easy to use. Thus takes Schrödinger's wave mechanics to hand to calculate the behavior of an electron when it collides with an atom. "Collision" and "wave"—isn't there a contradiction? Can waves collide with each other? The very word "collision" implies particles that collide with one another. No way to imagine it any other way. But if you describe an electron as a wave and it hits an atom, what could possibly result other than a scattered wave, a wave that goes in all directions. No need to even start calculating to predict this

---

[30] Born M., Mein Leben (1978, p. 310).

result, which Max Born indeed confirms. But even more interesting: if you allow a sharply localized wave packet to hit the atom (as an approximation for a particle), Born finds that the outgoing wave is not a cleanly localized wave packet, which in turn could be interpreted as a particle, but actually consists of individual waves.

But what, pray tell, is that supposed to mean? What is the significance of these waves? If you follow Schrödinger's naive notion, then the electron is a spatially localized wave packet before the collision. But after the collision, it is a charge distribution smeared in all directions according to the calculated wave pattern. Did the collision break the particle apart? Destroy it? Rob it of its wholeness? Or does the impact cause the particle to melt and become a smear-out charge distribution? Impossible! Electrons remain whole particles even after a collision with an atom; this is an experimental finding that no one can seriously doubt. Particles cannot decompose and spread across space.

So, indeed, what significance do these waves that Born calculated have? How do you interpret the waves that can be calculated using the Schrödinger equation? And here comes Born's interpretation, which represents his essentially own contribution to the new quantum mechanics. The waves that propagate from the location of the collision are not the particles themselves, but only—attention!—their probabilities.[31] Yes, indeed: Schrödinger waves are probability waves. This means: the calculations only provide a probability distribution. With a probability of 3% the electron will move in this direction after the collision, with a probability of 15% in that direction, and so on. The probability is high for wave crests and low for wave troughs. Quantum

---

[31] To be precise, the solution function of the Schrödinger equation must be multiplied by its complex conjugate function in order to determine the magnitude of the function, because only this magnitude can be interpreted as a probability.

theory cannot say in which of the many directions marked with percentage numbers the electron will actually move after the impact.

July 1926, Göttingen

"Quantum mechanics of collision processes." That is what Max Born called his quite brief preliminary communication, which he sent to the editorial team of the Journal of Physics on June 26th, followed by a more detailed article on July 21st. Schrödinger's wave equation only leads to probability waves. This interpretation by Max Born is in no small measure sensational. He is the first to recognize and express that these Schrödinger waves are not real at all. They are not mass distributions. They are not charge distributions. They are not reality at all. They will never be observable.

These waves are basically just a mathematical crutch that leads to a probability distribution. But if at the end of such a calculation, there are only probabilities for possible directions of the electron, how does one actually theoretically describe which path the electron actually takes? No one can describe this according to Max Born, because Mother Nature does not stipulate it. It is simply not "determined" at all. Only chance decides. There is no predetermined path for an electron. There is no determinism. Pure chance reigns specifically chance as part of Mother Nature. This then is indeed the end of the concept of a determined orbit for the electron, the same end that was Heisenberg's starting point back in Heligoland. In his second paper, Max Born writes an introduction that is beautifully clear and easy to understand:

1926, Antithesis: Wave Mechanics

The matrix form of quantum mechanics founded by Heisenberg and developed together with Jordan and the author of this communication is based on the idea that an exact representation of the processes in space and time is simply impossible and is therefore content with establishing relations between observable quantities that can only be interpreted as properties of movements in the classic limit case. Schrödinger, on the other hand, seems to attribute to the waves, which he sees as the carriers of atomic processes according to de Broglie's process, a reality of the same kind as that possessed by light waves; He tries to build wave groups that have relatively small dimensions in all directions and that are apparently intended to directly represent the moving corpuscle. Neither of these two views seems satisfactory to me. I would like to try to give a third interpretation here and test its usefulness on the impact processes. In doing so, I piggyback on a remark by Einstein about the relationship between wave field and light quanta; He said, roughly, that the waves only exist to show the way for the corpuscular light quanta, and in this sense he spoke of a 'ghost field'. This determines the probability that a quantum of light, the carrier of energy and momentum, will take a certain path; But the field itself has no energy or momentum.

With the complete analogy between light quantum and electron, one will think of formulating the laws of electron movement in a similar way. And here it makes sense to view the de Broglie Schrödinger waves as the ghost field or, better, "guiding field". The guiding field spreads according to Schrödinger's differential equation. But momentum and energy are transferred as if corpuscles (electrons) actually fly around. The orbits of these corpuscles are only determined to the extent that energy and momentum constrain them; beyond that, the pursuit of a specific orbit is determined only as a probability based on the value distribution of a function. This could be summarized, somewhat paradoxically, as: The movement of the particles follows probability laws, but the probability itself propagates in accordance with the causal law.[32]

Until Born's work, physics had believed that everything, really everything, that happens in nature has a reason, that is, that it follows a law, a causal law. If you know all the parameters appearing in the law at one point in time, the behavior of the system at a later point in time is absolutely bindingly determined by the law. This determinism was a sacred tenet of classical physics. There is always and everywhere cause and effect. With Born, however, the end of determinism has been reached. The electron is influenced by its guiding field, but then lets chance decide which of the possible paths it takes. Although there is an effect, there is no cause other than chance. And that is simply not a cause. A single electron does not reveal any law. Only when the paths of many thousands of electrons are superimposed does the resulting pattern show that a probability distribution has always been the discrete movie director in the background. Of course, it could be that this is not the end of the story, that there are still hidden parameters, and that a better, more profound theory will one day be invented that will restore determinacy. Born writes:

Of course, anyone who doesn't find comfort in this is at liberty to assume that there are other parameters that have not yet been introduced into the theory and that determine the individual event.[33]

Just to state his own position later:

---

[32] Born M., Quantenmechanik der Stoßvorgänge (1926, p. 803).
[33] Ibid. p. 826.

> *But I myself am inclined to give up determinism in the atomic world. But this is a philosophical question for which physical arguments are not the only determining factor.*[34]

And in this case the gentle Max Born was completely right, almost a prophet in the sense that only much, much later it became possible to establish beyond any doubt and with complete certainty that chance is not the defect of a provisional theory, but actually a genuine principle of nature. Almost a hundred years later, namely in 2022, the Nobel Prize in Physics was awarded for exactly this.

July 1926, Vienna

The receptionist was as polite as she was assertive. Doctor Adler is still in the midst of a difficult treatment that he cannot interrupt right now. Could he please take a seat and wait? It would take at least another half hour. She directed Pascual Jordan to a large uninviting leather armchair, which, despite its odd shape, turned out to be extremely comfortable. He reckoned it would be pleasant enough here to wait a good half an hour.

Pascual Jordan had an appointment with Dr. Alfred Adler, the famous psychotherapist from Vienna. Alfred Adler had dealt with nervous characters and that's why Max Born probably thought that he could also deal with Jordan's stutter. And so Born had organized an appointment with Alfred Adler for his protégé Pascual Jordan. Max Born didn't want to leave any stone unturned. He had been sending Jordan from one doctor's appointment to another all summer. This terrible stutter had to be treated somehow. Max Born, James Franck, and Niels Bohr pooled together a

---

[34] Ibid., p.866.

lot of money so that Pascual Jordan could go to the best doctors. Jordan visited three doctors in Bad Pyrmont alone, without success. And so now Vienna, to an appointment with Dr. Alfred Adler. Would he be able to help him?

On the long train ride to Vienna, Jordan studied the writings of Alfred Adler. With his "individual psychology," he was moving in a completely different direction than Sigmund Freud, who assumed in his theories that people are determined by their drives and traumas. Alfred Adler had a completely different opinion. For him, man was a completely free being who had to use his creative power to solve the tasks that life presented to him. Adler wanted to understand the psychological suffering of his patients through their life stories. Freud versus Adler. Ultimately, the question was: Are people determined or really free?

And that was a question that had interested Pascual Jordan since his early youth, for example when he was dealing with the materialism controversy of the nineteenth century. Are the results of modern natural sciences actually compatible with the idea of a soul, a God, and free will? Is man, as the materialists claim, ultimately determined in everything by his physical-chemical composition? Or is there, as the vitalists postulate, a *vis vitalis*, a special life force? Pascual Jordan had been a supporter of vitalistic biology for a long time, but then, inspired by the book *History of Materialism* by Friedrich Albert Lange, he converted to the materialistic point of view.

But for Jordan, the most interesting aspect of his train ride wasn't so much the debate between Adler and Freud and the parallels to the materialism controversy. Rather, Max Born's manuscript, given to him for his trip to Vienna, turned out to be not only the most riveting thing but also revealed a completely unexpected parallel to Alfred Adler's psychology. The manuscript was about Born's work on

1926, Antithesis: Wave Mechanics

collision processes including the probability interpretation of the wave function. According to Born, the new quantum mechanics ultimately only calculates probability distributions. Which of the various orbital possibilities an electron actually realizes is decided solely by chance?

This was an exciting realization for Jordan! Because chance took control, there was an effect, but no cause that brought it about. Effect without cause: The end of causality. The dynamics of the electron are ultimately not determined. *Not* determined! That hit Pascual Jordan like lightning. That physics would one day go so far as to show that in the end not everything is determined by natural processes. The debate between free will and determinism has not only been raging since the materialism dispute but has been an issue in philosophy at least since the ancient Greeks. And Born's interpretation of probability now opened up a whole new front. In the end, does quantum physics support the idea of free will by ruling out the idea of a deterministic universe? Doesn't the end of determinism mean the final victory of the idea of truly free will? What kind of insane inferences and interdisciplinary connections could still be anticipated? What, for example, will someone like Alfred Adler say when the news of the end of determinism reaches him? Pascual Jordan couldn't stop his awe. What a strange coincidence: learning something about the end of determinism from Born while on his way to Alfred Adler. Pascual Jordan had experienced a train journey that not only took him from Göttingen to Vienna, but also to the really big questions in life.

Would he please come? Dr. Adler will see him now. The resolute receptionist opened the door to the hallway for him. Over there, and then the second door on the right. And please keep your answers brief, Mr. Jordan, you only have one hour. My God, thought Pascual Jordan, I stutter, I can't give long answers. That's exactly the problem!

July 1926, Berlin

At the beginning of June, Erwin Schrödinger was still working on the last paper of his legendary series, the end of a 6-month tour de force through which he created wave mechanics. Already by the end of June his great travels began. He toured Europe and made himself and his theory known. This was clearly Erwin Schrödinger's summer, and everyone wanted to hear him. Including, of course, the great Berliners: Max von Laue, Nernst, Einstein, and Planck. It was Planck who invited him to give a lecture at the German Physical Society on July 16th and to the physics colloquium at the university on July 17th.

After the lectures, Schrödinger can hardly save himself from encouragement and pats on the back. Einstein believes that the work shows "genuine genius" and continues:

> *Not such an infernal machine, but a clear idea—and inevitable in its application.*[35]

After he had previously written to Schrödinger personally:

> *I am convinced that you have found real progress with your formulation of the quantum condition, just as I am convinced that the Heisenberg-Born path is misguided.*[36]

In contrast to the youngster Heisenberg with his Heligoland "infernal machine," the work of the Viennese

---

[35] Letter from Albert Einstein to Erwin Schrödinger dated April 16, 1926, and April 26, 1926 (Przibram, 1963, p. 21; 26).
[36] Ibid.

1926, Antithesis: Wave Mechanics 141

physicist was entirely to the taste of the Berlin old guard. In hand a differential equation that was compatible with de Broglie's obvious idea, familiar mathematics to solve these differential equations, beautifully clear wave functions as a solution, and finally the electron as a standing wave spatially arranged around the atomic nucleus. What more could one want? And since it was now proven by Schrödinger himself that Heisenberg's matrix mechanics on the one hand and Schrödinger's wave mechanics on the other were equivalent to each other, it was preferable to opt for the vivid Schrödinger picture and declare it to be the only true theory and push the Göttingen ideas to the sidelines. Max Born's wondrous and irritating conclusion that Schrödinger's waves cannot be real waves had not yet made the rounds in those days.

Nobody has ever found fame as quickly as Schrödinger. This must have been because the demanding and clumsy Göttingen matrix mechanics had put physicists in an unpleasant, even torturous state of tension, from which the wonderfully clear wave mechanics had beautifully released them. A rapidly spreading general relief began in science and it was linked to the name Schrödinger. Schrödinger = salvation. That was the association. Redemption from the abstract, intentional, difficult, unimaginable of the Göttingen hell machine.

For Schrödinger these must have been weeks of pure happiness. He must have been staggering in drunken bliss! The world-famous Albert Einstein, but especially admired by Schrödinger himself, spoke of his "genius." A shiver ran down Erwin's spine: the recognized genius calls him a genius! Of course, he had always known it. But who can bear such tremendous fortune? Only yesterday still a boring average theorist in Zurich and today a genius! And now, on the evening of July 16th, Einstein and Schrödinger stand close together at a small party hosted by Max Planck, both holding a glass of Riesling late harvest in their hands, putting their heads together and converse in a whisper. All eyes return again and again to the two of them. What are they saying? No one dares to approach them. Two true greats standing next to each other, so you have to keep a respectful distance. And Max Planck also sees the two of them standing there and—still impressed by Schrödinger's lecture—he suddenly thinks to himself: wouldn't this Erwin Schrödinger be ideally suited to succeed me as a professor here in Berlin? Can I get the ball rolling? Max Planck smiles.

July 1926, Munich

The Auditorium Maximum of the Ludwig Maximilian University in Munich. It's the July 23rd. What an imposing space! The ceiling: a barrel vault made of glass through which the sun's rays fall and illuminate the room as bright as day. Balcony on the left, balcony on the right, and at the top, almost sticking to the ceiling, plus another wide balcony that closes off the room at the back. Height and width, enough space for big thoughts. And everyone looks to the speaker's podium, where Erwin Schrödinger, 38, is looking through his papers for his lecture one last time. "Fundamental ideas of wave mechanics." The hall is packed and stuffy. Hopefully, they'll leave the doors open! How crowded it is. The *Gauverein of the Physical Society* can't have that many members! They said that only club members would be admitted. But these aren't all club members here, are they? But it makes sense. Everyone wants to see him. Erwin Schrödinger is finally in Munich. Why does he speak first in Berlin and then in Munich? Incomprehensible. How much longer before he starts? The rector of the University Wilhelm

Wien is standing at the front talking to whom? Can you recognize them? Yes, yes, that is of course Arnold Sommerfeld, head of the Institute for Theoretical Physics.

The usher is about to close the doors when a slim young man with straw-blond hair shoots through the door. Tanned, strong, boyish, and with the smooth movements of an athlete. It's Werner Heisenberg. He has traveled through half of Europe and has come straight from his hay fever vacation in Norway. Barely blasts through before the usher finally closes the door. His mentor and doctoral supervisor Arnold quickly shakes his hand. Heisenberg would have liked to greet Erwin Schrödinger as well, the man with whom he is nowadays invariably mentioned in the same breath, but whom he has never met in person. But there is no time left for that, because Rector Wilhelm Wien sits down and the club chairman begins his welcome speech. At least Heisenberg manages to give Schrödinger a quick nod, who also recognizes him and nods back. So here in Munich, on July 26, Schrödinger and Heisenberg meet for the first time. The two great combatants of quantum mechanics.

Messrs. Wien, Sommerfeld, and Heisenberg, on the other hand, have known each other for a long time and have already had the pleasure of each other's company, a particular and rather embarrassing pleasure that all three of them are now compelled to think about. It was exactly 3 years ago, in July 1923 at Heisenberg's *Rigorosum*, his oral doctoral defense. Wilhelm Wien is his examiner. Heisenberg was unable to answer a single one of Wilhelm Wien's exam questions. Not one. Just didn't prepare. Wien wanted to let him fail. However, Heisenberg's supervisor, Arnold Sommerfeld, who was deeply impressed by Heisenberg and his doctoral work, vigorously opposed this. You can't fail a Heisenberg. He is already far too famous for that. You would be making a fool of yourself. They argued, they really argued. They suffered from each other. A deeply embarrassing situation. But they finally agreed on the rating "cum laude." "Cum laude" for Werner Heisenberg's doctoral thesis, hard to believe, "cum laude" for an absolutely exceptional thesis and an absolutely exceptional physicist! None of the three could forget the experience anytime soon.

And now Erwin Schrödinger begins his lecture. A professor of theoretical physics at ETH Zurich. Absolutely confident, with a warm Viennese accent. And the mathematics is so beautiful of course. So easy, so understandable. It simply doesn't get any closer to the good old world of classical physics. Everything lines up so beautifully. The sun so bright; the room so light. The day is so lovely, the theory so clear. Everything will be fine; everything will be easy. And while the Austrian draws his listeners with every word more deeply under his spell, something is brewing inside Heisenberg and he has to hold himself back not to jump up and give a wild counter-speech. He sees the happiness in the eyes of the listeners and knows that he will not be able to win this fight. Let's listen to Tobias Hürter's wonderful description of the following scene:

> *Heisenberg was able to hold back—until the last moment, until the end of Schrödinger's second lecture. Then during the discussion that followed, the excitement bursts out of him. Quantum mechanics is his invention, and now Schrödinger stands as the man who saves the world from it? He can't let that stand. This is his city, his theory, his territory. This is where he went to school, studied and received his doctorate. Heisenberg raises himself and his voice. All eyes are on him. He vehemently speaks out against Schrödinger. How should atoms oscillate gently when so much experimental evidence shows that they do not? That abrupt collisions occur within them? What about the photoelectric effect, Franck collisions, Compton scattering? Schrödinger's theory cannot explain any of this without*

1926, Antithesis: Wave Mechanics 145

particles, without discontinuities, without quantum jumps— precisely without the things that Schrödinger wants to abolish.[37]

Very dramatic, but did Heisenberg really become as ferocious as Hürter would have you believe? True, however, that Wilhelm Wien is now angry. So perhaps Heisenberg must have been a little too forceful. Let's listen further:

An angry Wilhelm Wien jumps up and reprimands Heisenberg. He understands Heisenberg's agitation that Schrödinger had done away with matrix mechanics, that quantum jumps, and all that nonsense are now over. Professor Schrödinger will surely answer all remaining questions soon. Now Wien finally sees his opportunity to fail Heisenberg after all. "Young man, you still have to learn physics," he counsels Heisenberg and gestures to him to sit down again. He almost kicked me out of the lecture hall Heisenberg would later say.[38]

What might the audience have thought? Did anyone suspect that they were witnessing two future Nobel Prize winners sparring with each other? In essence they are both correct, that there are waves, but only probability waves, and that on the other hand there are particles, there are electrons that follow their own, unpredictable path and can jump back and forth between different states. If only Max Born had been there to pacify the two cock fighters with his clear-sighted interpretation. Here in Munich, in July 1926, something began, something that Einstein and Bohr hotly debated well into the 1930s: namely the question of what exactly was grasped with this new theory and how to actually implement it. "Young man, you still have to learn physics!" This arrogant sentence from Wilhelm Wien was meant differently but wasn't all that wrong. Not just Heisenberg,

---

[37] Hürter (2021, p. 200).
[38] Ibid.

actually all physicists still had to learn physics; have to learn physics from scratch. It's only now that things get interesting, when we begin to ask ourselves: What do we actually learn from all of this?

Arnold Sommerfeld remains politely silent the entire time and looks down. When he says goodbye, however, he gently pats his student, the defeated and dejected Heisenberg, on the shoulders. It'll be okay, boy, it'll be okay! Just don't let it get you down!

July 1926, Munich

Heisenberg is on vacation with his parents in Munich. The experience at the LMU Munich still haunts him. He is extremely angry about Wilhelm Wien's reprimand. And even more about the fact that Schrödinger is so successful with his theory. What does a Heisenberg do when he is extremely angry? He writes to Wolfgang Pauli. On July 28th:

Dear Pauli,

Thank you very much for your beautiful book, which I read critically and unforgivingly, but with a lot of pleasure. It is an exact representation of the physical connections that were known before the chaos of last year, and reading it was a real relief for me after Schördinger's lectures here in Munich. As nice as Schrödinger is personally, I find his physics puzzling: when you hear it, you feel 26 years younger. Schrödinger throws everything "quantum theory", namely the photoelectric effect, Franck collisions, Stern-Gerlach effect, etc. overboard; this makes it easy to create a theory. But it just doesn't line up with experience.[39]

---

[39] Letter from Heisenberg to Pauli dated July 28, 192, see Klein and Toomer (1979, p. 337).

The end of this letter from Munich is also interesting. He reports on his work with the helium atom that he did during his stay in Norway's mountains due to hay fever. Key phrase: driving a steam roller, as he described it in a letter to Pauli in June.

I have now sent off my work on the helium spectrum with feelings of doubt and am not really satisfied. The calculations are all so imprecise and incomplete, the nicest thing is the fine structure that turned out correctly. In any case, there is still a lot of work to be done in terms of quantitative agreement. Now thank you again for the beautiful book, say hello to Hamburg physics and don't work too much during the holidays. Congratulations on the professorship, Werner Heisenberg.[40]

The *Zeitschrift für Physik* actually confirms receipt of a Heisenberg article for July 24th: "On the spectra of atomic systems with two electrons. By W. Heisenberg in Copenhagen." What is extremely interesting is that a large part of the calculations in this work are based on Erwin Schrödinger's hydrogen wave functions. So while he loudly campaigns against Schrödinger's interpretations, he—by continuing to use them—quietly and secretly recognizes the value of Schrödinger's work.

August 1926, Cambridge

Every physicist's heart now beats for Schrödinger. And Heisenberg's matrix mechanics were on the defensive. Barely in the world and already an outcast? No, thinks Heisenberg, we have to counter this. And so Heisenberg plays his last card: Paul Dirac.

---

[40] Ibid. p. 338.

I think I have to quit physics. A young Englishman came along, his name is Dirac, he's so clever—it's hopeless to compete with him.[41]

This is how Heisenberg speaks of Dirac and then probably says to himself: "If you can't beat him, join him." And so he joins Dirac. Asks him to help him and finally tackle this strange Schrödinger equation. And the quiet genius has mercy. Dirac sits down at his desk, rubs his hands, and looks at Schrödinger's papers again, which he had angrily put aside in February:

I think I did read Zeitschrift für Physik quite a lot. That was the main journal in those days. Yes, I read the Zeitschrift für Physik quite a lot. I had learned German at school. Not fluently, but enough to be able to read scientific German at that time.[42]

Paul Dirac is right. In fact, the *Zeitschrift für Physik* was the leading journal for physics research not only in Germany but worldwide in the 1920s. This is all the more astonishing considering that it was only founded in 1919 by, among others, Albert Einstein and Fritz Haber. Especially in these years, newspapers and magazines received a huge boost from a small revolution in printing technology: high-speed printing presses. Rotary presses had been around since the 1890s, but by 1927 the technology had advanced to the point where 100,000 newspapers could be printed per hour. The mass production of printed matter created a tsunami in the social and political world of the time. The world became smaller as new magazines like *Time* and *Reader's Digest* flooded society and quickly gained millions of readers. With the advent of paperbacks at the beginning of the

---

[41] Rößler (2009, p. 259).
[42] Dirac P. (1962).

twentieth century, sales of literature exploded and reached entirely new target groups. Once reserved for the privileged, reading has now become affordable and accessible to all. And in science, too, the mass-produced scientific magazines revolutionized reading behavior. Journals crossed borders in larger quantities and cross-border scientific communities and international research teams were formed almost automatically. New printing techniques certainly played their part in the emergence of quantum mechanics. Without this medium, how could Cambridge-based Paul Dirac have known what his contemporary Pascual Jordan in Göttingen, what Schrödinger in Zurich, or Heisenberg in Copenhagen were doing and thinking?

But back to August 1926. It was a blazing hot summer in Cambridge. The college was deserted. A deep summer silence. A silence that Paul Dirac had dreamed of throughout the turbulent spring. No, not a vacation like his mother had recommended. No, no relaxation. No, not to Bristol. Just sitting in your own room in college. That was his plan for the summer. Dirac, who doesn't drink, hardly eats, and rarely sees friends if he has any at all, who prefers to be by himself. Dirac is happy. He loves nothing more than a regular, even monotonous life. A quiet, lonely summer in Cambridge was the ultimate fulfillment for him: five days of theories from morning to evening, then on Saturday the more technical problems in his calculations and finally on Sunday: hiking. For him, that was the true rhythm of his life. No distractions whatsoever. Like a monk in prayer: just concentrating on his work. Just concentrating on what is *really* important. And always at work on something great and lasting.

This was the moment when Dirac decided to allow Schrödinger's wave functions to enter the crystal palace of his mind. And how this happens can hardly be put into words. Let's go through his paper. He first explains Heisenberg in two sentences. Then Schrödinger in two sentences and in both cases you think: boy, he sure summed this up succinctly and to the point. Then he starts up again. Writes down an energy expression as if still in classical physics, introduces differential operators for the momentum and the energy (you might be surprised), inserts these into the classical expression, now adds a wave function and with these few simple steps gets back to the Schrödinger equation as if the entire process had been a piece of cake. It is, of course, just a different representation of the same thing. Schrödinger lets a finished differential equation, Dirac, on the other hand, just lets operators fall from the sky. Operators that, when inserted into a classical expression, lead to Schrödinger. But these operators allow you to immediately understand, for example, why the position-momentum commutation relation is the way it is. In any event, Dirac proceeds to slowly and skillfully lead the entire formalism toward the matrices in the Heisenberg representation. Formally, he once again showed the equivalence of Schrödinger's wave mechanics and Heisenberg's matrix mechanics, just as Schrödinger or Pauli or Carl Henry Eckart had done before him.

But he achieved it by introducing energy and momentum operators. And these operators can be found in every representation of quantum mechanics today. John von Neumann, who is the same age and who, incidentally, arrived in Göttingen as a young student this very spring to learn quantum mechanics from David Hilbert and Max Born, will later be the author of a strictly mathematically thought-out book on quantum mechanics, in which these very operators played a key role. This book enjoyed a unique reputation but was outdone by the legendary book by Paul Dirac "Principles of Quantum Mechanics," which was published in 1930 and, of course, introduced quantum mechanics via operators, via those same operators that first appeared in writing in that hot summer in Cambridge.

"On the theory of quantum mechanics" is the name of the article in question here and which Dirac will submit to the Royal Society on August 26th. It reads like a modern textbook on quantum mechanics: something is already finished and completed here, the spin is still missing, many details are still missing, but he has artfully interwoven the two main strands of development, Heisenberg and Schrödinger. Dirac, who was completely unknown a year ago and completely clueless about quantum mechanics, already appears in this paper as an old master: confident and calm. Someone who understands everything deeply and thoroughly. Someone who bows down to us humans and explains it to us so simply, so lovingly, so naturally, as if he were talking about something he has been thinking about his entire life. Freeman Dyson puts it this way:

> *The great works of the other quantum pioneers were more frayed, less perfectly formulated than Dirac's. His great discoveries were like intricately carved marble statues falling from the sky, one after another. He seemed capable of conjuring up natural laws through pure thought—it was this purity that made him unique.*[43]

By now Dirac should have understood that fate is smiling on him and has a task in store for him that matches his abilities. The mathematical formulation—and indeed the final formulation—of quantum mechanics: that is his big task. And now, in the summer of 1926, thanks to this ability he leaves his last pursuers behind and takes a commanding lead in the field.

---

[43] Interview with Freeman Dyson, June 27, 2005 (Farmelo, 2016, p. 459).

But the old master Dirac brings together not only the works of Schrödinger and Heisenberg in this momentous paper "On the theory of quantum mechanics." That covers only two of five chapters. He also tackles the famous paper by Wolfgang Pauli: the exclusion principle, that Pauli had discovered at the beginning of 1925. He looks at systems with many electrons. For example, the electrons in heavier atoms. Now each electron is given its own wave function. Let's call this wave function for the single electron an orbital. Dirac now combines all of these orbitals to form an overall wave function for the system made up of all electrons. For this construction, there are basically only two ways in which the overall wave function can respond to the exchange of two electrons. If it doesn't react at all and stays as it is, then we speak of a symmetric overall wave function; or it gets a negative sign when electrons are exchanged, in which case it is an anti-symmetric overall wave function. So far so good. But now you can immediately deduce from Dirac's representation what happens if you give two electrons the same orbital in the symmetric or anti-symmetric overall wave function. In other words, they are placed into the same quantum state. Nothing happens with the symmetrical wave function. It can handle this. The antisymmetric wave function, on the other hand, disappears. Is zero. It no longer exists. And thus, he arrives back at the Pauli principle, which Dirac can now express in a much simpler fashion: Electrons in a multi-electron system always have an antisymmetric wave function, whose property automatically ensures that Pauli's law is followed. Just semantics, one wonders. No. On the contrary. Something fundamental. Mother Nature contains elementary particles with symmetric overall wave functions as well as those with anti-symmetric overall wave functions. Physicists call the two classes: bosons and fermions. Like men and women. Two different genders, so to speak. Electrons are fermions. Photons are bosons. While electrons obey the exclusion principle because of this property, photons, and all other bosons, can always occupy the same state together. And finally, Dirac uses these considerations and now easily derives the statistical distribution for gas of free electrons, a distribution that Enrico Fermi and Pascual Jordan had already found and which is now called the "Fermi-Dirac distribution."

All of this in Paul Dirac's wonderful paper "On the theory of quantum mechanics," written on a single hot summer in Cambridge, a summer a normal person likely would have whiled away quietly and dreamily in the cool shade of a tree.

September 1926, Berlin

High up and bathed in the blue Berlin sky,
 A steel tower is built.
 Steep into the Berlin air,
 Illuminated by the last scent of summer.[44]

Hans Bredow recites these dubious rhymes in front of over 1000 invited guests and dignitaries. Some in the audience must have felt embarrassed for him. Hans Bredow is an engineer and chairman of the Reich Broadcasting Society and stands at the foot of the 146-meter-high Berlin radio tower, which will finally be inaugurated on September 26. The Berliners call it *der Lange Lulatsch*: The lanky lad. Bredow had already experimented with transmitting radio broadcasts during the First World War, and the first news

---

[44] "Berliner Funkturm." In: Wikipedia—Die freie Enzyklopädie. Bearbeitungsstand: November 12, 2023, URL https://de.wikipedia.org/w/index.php?title=Berliner_Funkturm&oldid=239048357

was sent from the post office in Berlin as early as 1920. In those days, radio simply meant the host alternated between reading a text and putting on a record. However, the enormous potential of the idea quickly became apparent, and the new technology was used in a variety of ways. Music and entertainment programs, news programs, interviews, and football commentary were broadcast. As a result, the need for large-scale infrastructure grew and so in 1926 plans were drawn up for a radio tower that would subsequently shape the Berlin skyline. It was officially opened at the third "Great German Radio Exhibition" and housed a restaurant 50 meters up in the air that is still in operation today! The new monument welcomed hundreds of thousands of visitors in the first few years after its opening and became a symbol of the exciting new spirit of modern Berlin and the wonders of technology.

September 1926, Göttingen

"Oh my darling, oh my darling, oh my darling Clementine!" It sounds loudly through the train station in Göttingen. This is definitely not a radio. No, that's Howard Percy Robertson! The American! So he's back again! Ah yes! There he is! He appears to be picking up a friend from the train station. Wherever Robertson shows up, he starts to sing. All of Göttingen already knows him. And he always sings the same tune. Namely: "Oh my darling Clementine!" And many people bellow along. Howard, later a professor at Caltech and an employee at the CIA, is currently doing his postdoc in Göttingen. The all-present international atmosphere here gets him going every time and he cannot help but channel his enthusiasm through obsessive singing.

1926, Antithesis: Wave Mechanics

Göttingen was suddenly the center of the world. After Max Born's tour of the USA last winter, American students came to Göttingen in droves. And not just Americans. Waves of visitors from abroad poured into Göttingen. Mainly because of quantum mechanics. September was the peak season for science tourists from all over the world. But these weren't normal tourists. They were touring geniuses, child prodigies, clever fools, wild people, young people from all over the world, among whom Oh-my-darling-Robertson was the most normal. They were all overflowing with ideas. Max Delbrück, a recent student of Max Born, summed it up nicely:

The guests in Göttingen were of extremely bizarre natures, brilliant lunatics, unworldly and decidedly difficult in dealing with their fellow human beings.[45]

Of course, quantum mechanics was also done in Copenhagen and Cambridge. But everyone talked only about Göttingen. And while Niels Bohr and Werner Heisenberg in Copenhagen quietly and out of the limelight started to tackle the final chapter on the origins of quantum mechanics, the world again and again only had eyes for Göttingen as the mecca of quantum mechanics. Augustus Trowbridge, responsible for assessing the quality of European physics faculties for the Rockefeller Foundation and its scholarship recipients, came to the conclusion regarding physics in Göttingen:

It's probably the best thing I've seen on the continent.[46]

---

[45] Greenspan (2006, p. 151).
[46] Ibid.

And that had a lot to do with Max Born and his work. It makes your heart clench when you recall how the Nazis in 1933 brutally and deliberately destroyed this unique atmosphere in Göttingen within just a few weeks by depriving the most capable and successful scientists of their professorships. Incomprehensible! "You are lost and gone forever, oh my darling Clementine!"

September 1926, Copenhagen

At some point in the summer, Dirac came up with the remarkable idea of leaving the British Isles and having a look around the continent. This is what we call post-doc time today. Actually, he just wanted to go to Göttingen. But his supervisor Ralph Fowler had insisted that he also visit Niels Bohr in Copenhagen, a suggestion that Dirac agreed to only reluctantly and only because Heisenberg, who was the same age, was now an assistant there. Finally, on September 8th, the journey by a ship across the North Sea to Denmark. High seas and then a 16-hour drive, and Paul Dirac struggled with motion sickness almost the entire time. Was this perhaps an omen for his collaboration with Bohr?

Bohr and Dirac could not have been more different. One didn't speak at all or actually only spoke the three words "yes," "no," and "no idea." And the other followed his principle of "never speak more clearly than you think." And Bohr didn't think so clearly. Bohr loved to say unfinished things. To talk in pictures. To express himself in a completely confused way, gesticulating wildly. Bohr wanted to use words to understand things, whereas Dirac wanted to use mathematical formalisms. Bohr didn't like and couldn't do mathematics; Dirac didn't like words. Bohr loved the interpretation of derived equations; Dirac considered it unnecessary. The two of them marveled at each other, surely shaking their heads in their imagination.

Dirac:

Bohr's arguments were mostly qualitative in nature and I couldn't really make out the facts behind them.[47]

Bohr:

This Dirac seems to understand a lot of physics but doesn't say a word!

Dirac:

What I wanted were statements that could be expressed in equations, and Bohr's work very rarely provided such statements.

Bohr:

The strangest person who ever visited my institute in Copenhagen.

For a comedian, the couple would have provided much grist for the mill. Things went wrong right from the start, even in the first week. Bohr often asked younger colleagues to assist him in writing articles. He dictated and you had to follow verbatim; an absolutely torturous procedure given that Bohr couldn't speak straight. Back and forth, edit, cut, oh no, changed my mind, start anew, messed up, sorry, change it after all, no, completely new, shorten, everything once more, bit by bit, confused and unclear and you never got to the end! And now Paul Dirac. In his first week as a writing assistant to Bohr, Dirac stops after two sentences

---

[47] This and the following three quotes can all be found at Hürter (2021, p. 204).

but doesn't say anything. Bohr continues speaking. Dirac looks. Bohr speaks. After a few minutes, Bohr notices that Dirac is not writing. Why isn't he writing? Was he too fast? No. Too slow? No. Then why doesn't he write? Finally, Dirac:

> *In school I was always taught not to start a sentence until I knew how to end it.*[48]

Bohr looks at him with wide eyes. He keeps dictating and Dirac continues to look at him. And finally gets up, puts the blank piece of paper on the desk, and walks out. His employment as Bohr's writing assistant lasted about half an hour and was probably the most fruitful period of collaboration between these two great physicists.

Another attempt by Bohr to relate to Dirac and win him over was also a legendary failure. A walk to Kronborg Castle, which as the setting for Shakespeare's *Hamlet* surely must interest an Englishman. He talked and talked and talked and talked to the silent Dirac, who was walking next to him visibly uncomfortable. Finally, Dirac cut him off and explained curtly that he was not interested in literature or language and preferred the purity of mathematics as a means of expression. While Bohr probably experienced Dirac as uptight and even a little messed up, Dirac saw Bohr as an eccentric. In later years, Dirac recalled Bohr's peculiar way of designing experiments to test psychological theories:

> *He did not concern himself with metaphysical problems. But he concerned himself with problems which are not related to science at all. For instance, when two gunmen each draw a pistol and each point it at the other, each one wants to kill the other one, but no one dares to shoot. What's the explanation for that?*

---

[48] Farmelo (2016, p. 113).

1926, Antithesis: Wave Mechanics

Why doesn't one of them shoot? Bohr bought some toy pistols and tried this out with various people in the lab.[49]

And yet Dirac flourished at Bohr's institute. It was probably this informal and stimulating atmosphere that Bohr created around himself. The casual interactions, where you could let off some steam and then continue working together afterward. All this was a world away from the traditional and strictly hierarchical world of Cambridge. While in Cambridge efforts were made to maintain monastic silence and austerity, the young researchers in Copenhagen were encouraged to have fun together outside of their work, drinking beer, going dancing, getting together in bars or even playing table tennis with each other in the library. Bohr himself was inconceivably famous in his own country, and his multidimensional nature and his eventful and rich life outside of science made a deep impression on the young Dirac. He got along well with the other postdocs, with Werner Heisenberg, for example, in Dirac's eyes an open, accessible, and generous colleague. On the other hand, he couldn't stand Wolfgang Pauli at all because of his arrogance and brutal directness. And what did Dirac's colleagues think of him? To Christian Möller, a young Danish physicist at the institute, Dirac appeared distant:

He often sat alone in the innermost room of the library in the most uncomfortable position, so absorbed in his thoughts that we hardly dared sneak into the room for fear of disturbing him. He could spend all day in this same position and write an entire article, slowly and without crossing anything out.[50]

---

[49] Dirac P. (1963).
[50] Møller, C. (1963) "Some memories from life at Bohr's Institute in the late 1920s," in Niels Bohr, et Mindeskrift [Niels Bohr, a Memorial Volume], Copenhagen: Gjellerup, pp. 54–64, see Rechenberg (2010, p. 540).

In this library, remaining in the most uncomfortable position imaginable, Dirac used to "balance on a dizzying path between genius and madness," in the words of Albert Einstein. He had a reputation for being quite aloof. In the evenings, while others were relaxing with a beer or a movie, Dirac would take long walks through the streets of Copenhagen, sometimes taking the tram to the last stop and walking back to his accommodation.

October 1926, Copenhagen

The Pope summons you to an audience in Rome. Pope Bohr had of course gotten wind of the scandal in the LMU auditorium between his assistant Heisenberg and Erwin Schrödinger. And whoever messes with his assistant also messes with him. And so he invited Schrödinger to Copenhagen in order "to be able to delve deeper into the open questions of atomic theory in the inner circle of those who work here at the institute." Copenhagen was indeed the stronghold of quantum mechanics at the time: Heisenberg was now there as an assistant, Dirac was doing his post-doc, and even Max Born's assistant, Friedrich Hund, happened to be in Copenhagen. These were the boys, all in their early twenties. Schrödinger and Bohr, on the other hand, were the old timers, even though they weren't really all that old: Schrödinger 39 years old, Bohr 41 years old. The two of them had never met.

They would meet for the first time today, October 1, 1926, when Erwin Schrödinger gets off the train in Copenhagen and is greeted by Niels Bohr. Schrödinger will stay in Copenhagen for 8 days, give a lecture to the Physical Society on the fourth day, but will otherwise be a guest in the house of Niels and Margrethe Bohr. It's going to be a

1926, Antithesis: Wave Mechanics

really arduous time for poor Schrödinger. The two combatants Bohr and Schrödinger essentially debate and argue the whole time. Tirelessly and with unabating intensity. And neither yielding even a millimeter of ground to the other. Schrödinger represents his view that his waves are real, that the electron is a smeared charge distribution in the atom, that it does not make hard quantum jumps and that there are no discontinuities. For him, quantum physics is straightforward and directly linked to classical physics. For Niels Bohr, however, quantum physics is a radical break with all classical views. He finds the idea of a smeared charge distribution intolerable. Despite the wave function, the electron always remains a particle. And this particle simply jumps.

Heisenberg tends to hold back in these discussions; the Munich experience still lingers within him. In his autobiography "The Part and the Whole" he reports on these eight legendary days:

> *The discussions between Bohr and Schrödinger began at the Copenhagen train station and continued every day from early morning until late at night. Schrödinger stayed in the Bohrs' house, so there could hardly have been an interruption in the conversations for external reasons. And although Bohr was otherwise particularly considerate and kind in his dealings with people, here he seemed to me almost like an implacable fanatic who was not willing to make even the smallest concession or allow even the slightest doubt. It will hardly be possible to convey how passionate the discussions were on both sides, how deeply rooted the convictions were that could be felt behind the sentences spoken by both Bohr and Schrödinger. So what follows can only be a very pale reflection of those conversations in which they fought with utmost energy about the interpretation of the newly realized mathematical representation of nature.*[51]

---

[51] Heisenberg W., Der Teil und das Ganze (2022, p. 92).

It is impossible for the gripping discussion that Heisenberg describes so beautifully in his autobiography to be reproduced here in its entirety. But the end of his description needs to be quoted:

> *The discussion went on for many hours of the day and night without reaching any agreement. After a few days, Schrödinger became ill, perhaps as a result of the enormous exertion; he had to stay in bed with a feverish cold. Mrs. Bohr looked after him and brought tea and cake, but Niels Bohr sat on the edge of the bed and spoke to Schrödinger: "But you have to acknowledge that . . . ."*[52]

Just imagine: the two have only known each other for a few days. And the big and heavy Niels Bohr is sitting on the edge of Schrödinger's bed, who is too ill to sit up! Bohr in a tie and Schrödinger in his pajamas. And Bohr trying with unrelenting intensity to persuade Schrödinger, who in his febrile state can hardly withstand the massive onslaught! No, Schrödinger indeed was not in an enviable position. He later wrote about these discussions:

> *Bohr's attitude towards the nuclear problems is really peculiar. He is completely convinced that understanding in the ordinary sense of the word is impossible. The conversation almost always immediately heads towards philosophical questions and you soon no longer know whether you really take the position that he is fighting and whether you really have to fight the position that he is taking. Certainly the point of view of vivid images that de Broglie and I adopt is not nearly advanced enough to give an account of even the most important facts. And it is outright likely that here and there we have taken a wrong path that must be abandoned.*[53]

---

[52] Ibid. p. 94.

[53] Letter from Schrödinger to W. Vienna, October 21, 1926, see Klein and Toomer (1979, p. 339).

Ah-ha, Schrödinger hints at compromise! In contrast to the Copenhagen group which was "very sure towards the end of the visit" that they were on the right track. If Heisenberg's stay in Heligoland completed the first stage in the history of the emergence of quantum mechanics and if Schrödinger's Arosa vacation perhaps represents the end of the second, then this visit by Schrödinger to Copenhagen marks the beginning of the third phase: the phase of the "Copenhagen interpretation" of quantum theory, which was to come to a provisional end at the Solvay Conference a year later.

October 1926, Copenhagen

Oh, Paul Dirac wonders, a letter from Rome. Who could that be? He closes the door because Niels Bohr in his typical manner is rushing through the institute shouting loudly. He leans back in his comfortable desk chair, opens the envelope expectantly, and pulls out a typed letter from the Italian physicist Enrico Fermi:

> *In your interesting paper 'On the theory of Quantum Mechanics' you have put forward a theory of the Ideal Gas based on Pauli's Exclusion Principle. Now a theory of the ideal gas that is practically identical to yours was published by me at the beginning of 1926. Since I suppose that you have not seen my paper, I beg to attract your attention on it.*[54]

Ouch! That was unpleasant! "On the theory of Quantum Mechanics" was his miracle paper from the hot summer in Cambridge. Publishing the same result at the same time as another is no one's fault. But publishing something that has

---

[54] Letter from Enrico Fermi to Dirac, Dirac Papers, 2/1/3, Florida State University Libraries, Tallahassee, Florida. see Farmelo (2016, p. 106).

already been published a good 6 months prior is extremely embarrassing. Did Dirac simply not notice? No, just the opposite. He had even read Enrico Fermi's paper:

> *When I looked through Fermi's paper, I remembered that I had seen it previously, but I had completely forgotten it. I am afraid it is a failing of mine that my memory is not very good and something is likely to slip out of my mind completely, if at the time I do not see how it could be important for any of the basic problems of quantum theory; it was so much a detached piece of work. It had completely slipped out of my mind, and when I wrote up my work on the asymmetric wave functions, I had no recollection of it at all.*[55]

The miracle paper contained several results, and this was just one of them. And Dirac had derived the result in a completely different way from Fermi. But a shadow is cast over Dirac nonetheless. He habitually read his colleagues' papers—and if the results didn't impress him, he checked them off mentally and pushed them so far out of his mind that he didn't remember them even when he derived the same result at a later point. Dirac well may have been on the autism spectrum.

October 1926, Copenhagen

Friedrich Hund, Paul Dirac, and Werner Heisenberg often sit together. Usually right after lunch, when the stomach is still heavy and you can't go back to work yet. The three of them like to think of Betty Schultz, Niels Bohr's secretary. Because of the coffee. Because Betty Schultz makes truly respectable coffee. And so the three boys enter Betty's office

---

[55] Schwartz (2017, p. 108).

and walk past her desk with a friendly greeting. Betty then often moves her reading glasses down her nose and peers over them at the completely dissimilar trio rather mockingly: "Unbelievable!" her expression wants to say: "These three greenhorns are supposed to be great physicists? Bohr can't be serious." Now the three of them get their midday coffee from the table next to the sink and curl up with it in the institute library because the armchairs there are so soft and deep. They sprawl out on the well-worn seats and start to chitchat. Almost always about quantum physics. Of course, Dirac doesn't chat, Dirac just looks, but at least he looks quite friendly and as if he may even be enjoying the gathering. But quickly it's back to their desks; after all, all three want to create something of lasting value.

In these October weeks, however, they often linger in their armchairs longer because a letter is making the rounds, a letter from Wolfgang Pauli, which all three of them again and again want to clutch and are reluctant to hand over.

Dear Pauli!
Thank you for your long letter. I am replying with such delay because your letter is making the rounds here and Bohr, Dirac and Hund are fighting over it.[56]

Thus writes Heisenberg to Wolfgang Pauli at the end of October. So what is in this long letter from Pauli that makes people fight to read it? It came out to 14 pages, written on the evening of October 19th, and Pauli only put his pen down by the time the next day was already 3 hours old. He writes about Fermi-Dirac statistics, he talks about undigested dumplings and is apparently digesting the Max Born's papers regarding the quantum mechanics of an

---

[56] Letter from Heisenberg to Pauli dated October 28, 1926, see Klein and Toomer (1979, p. 349).

electron–atom collision which he makes plausible in his own way. He looks at the quantum mechanical rotator and wonders why it sometimes moves to the left, sometimes to the right, and when it changes direction. Then, in the spirit of Max Born, he interprets Schrödinger's wave function as a probability density in spatial space.

And finally, he cracks a joke, a "mathematical joke," as he writes, but a joke that is not at all funny. He formulates Schrödinger's wave function, but instead of in local space, now also in momentum space. Because he finds it somehow amusing to describe quantum mechanics with position-dependent wave functions and with momentum-dependent wave functions. So to speak, a q-representation and a p-representation of the wave function. For fun. What a grandiose joke! But Paul Dirac perks up at this mathematical joke and finds this transformation from one representation to the other representation incredibly exciting, and based on this he develops the transformation theory, which will see the light of day in December. So at least one person laughed at Pauli's joke. What a prankster, this Pauli! Just as other people write about a trip or talk about a show, Pauli writes about quantum physics. And you can tell that he is allowing himself the occasional sip of red wine while he is easily writing. It's absolutely fascinating how anyone can write about these challenging things so effortlessly. After a lot of mathematics, physics gets its turn:

> *So much for mathematics. The physics of this is still largely unclear to me. The first question is why only the p's, and at least not both the p's and the q's, may be prescribed with any precision.*[57]

---

[57] Letter from Pauli to Heisenberg dated October 19, 1926, see Klein and Toomer (1979, p. 346).

And at this point Heisenberg will have perked up because there is now a direct line from this letter to Heisenberg's uncertainty principle, which will arise weeks later. The q's are the location coordinates. And the p's are the momentum coordinates. In classical physics, a particle can be assigned three position coordinates and three momentum coordinates at the same time. Makes a total of six numbers, which then determine the position and momentum of a particle. This no longer works in quantum mechanics, where there are no longer six numbers, but only three. Hmm? What is going on? Wolfgang Pauli states:

One cannot assign ordinary c-numbers to the p-numbers and the q-numbers at the same time. You can look at the world with the p-eye and you can look at it with the q-eye, but if you try to open both eyes at the same time, you will go crazy.[58]

This means: in quantum mechanics, the electron is described by a wave function, which can be viewed as a function of three spatial coordinates or as a function of three momentum coordinates, but in any case only as a function of three coordinates. You can look at quantum mechanics in the q-representation or in the p-representation, but you can by no means determine the q's and the p's together and with any desired accuracy. If you know the place, the impulse is somehow gone. If you know the impulse, there is no place anymore. Wolfgang Pauli's letter, written on the night of October 19th, can easily be viewed as preparatory work for the uncertainty principle that Heisenberg clarified in the following months. And if you study Heisenberg's reply letter closely, you can sense how much this letter must have inspired him and how the wheels in his brain started turning. And work started just the same in Dirac's head.

---

[58] Ibid. p. 347.

But of course, he doesn't say anything and he certainly doesn't write letters.

October 1926, Göttingen

Robert Oppenheimer enters the scene. 22 years old. Born and raised in New York, later to become famous as head of the Manhattan Project. Max Born had met Oppenheimer in Cambridge and invited him to Göttingen, where he could do his doctorate with him. In Göttingen, Oppenheimer was generally considered a person who was difficult to get along with: "extremely nervous, tense, uptight and coming on strong," according to one of Born's colleagues. Born himself later said of Oppenheimer that "he was aware of his superiority in a way that was embarrassing." Yet he had good manners, was extraordinarily well educated, and incredibly perceptive. But unfortunately also unbearably arrogant. He judged much of what was said, commented upon, and discussed in the Born group "trivial" and reacted to it with almost devastating condescension. He worked with Born on the paper "On the quantum theory of molecules," which resulted in the so-called Born-Oppenheimer approximation. In the Born seminar, he habitually rudely interrupted the lecturers by approaching the blackboard, grabbing some chalk and starting to pontificate: "This can be solved much better in the following way." He had a way of stuttering that he himself perceived as elegant and apparently was fake, which is why Pauli used to call him the "Hm-hm-hm man."

November 1926, Berlin

The nominating committee to appoint Max Planck's successor at the University of Berlin has already been meeting for 3 hours. It's November 2nd. The chairs are hard, and everyone is longing for the end of the meeting. And given that they have been at it since the summer and have finally agreed on a list of three candidates after much back and forth, the meeting can now conceivably actually come to an end.

It is clear to all members of the commission: the three names on this list are the three greatest theoretical physicists in the world. Because only the greatest of the greats are allowed to follow an exceptional scientist like Max Planck. But who is on the list? Arnold Sommerfeld from Munich in first place. 58 years old, only a few more years until retirement. Firmly established in Munich. He's not going to come; this is just a small nod to the life's work of a great man. That's why there was hardly any discussion about the first place. The commission fought a lot more about filling the second place on the list. Well, it was a bit of a haggling before consensus was reached. But now, thank God, it was settled! Max Born took second place. The name Max Born, which also resonates so widely in America, his work on quantum mechanics, the great atmosphere that he created around him in Göttingen, that was the deciding factor. And then finally in third place on the list was Erwin Schrödinger, who essentially had entered the stage only six months prior. So second place is Göttingen matrix mechanics and third place is Vienna wave mechanics. That was the fight behind closed doors.

Fritz Haber finally requests that the meeting be adjourned. People are generally relieved. Finally done. Sommerfeld will decline and Max Born will be his successor. Now Walther Nernst—the "Nernst of life"—rises heavily and slowly from his seat and looks over at his colleague Max Planck. No, he does not agree to adjourn the meeting, not at all. Ultimately, it is about the successor to the honorable Max Planck and he deserves to be heard again. Max Planck clears his throat. He was usually silent in meetings, preferring to let the others fight to exhaustion. And just when no one can go on and the chairs have become too uncomfortable, he would speak up.

And that is what he did now. He made the request to swap the names in second and third places on the list. Schrödinger to second place and Born to third. The faculty later reports to the ministry, what Planck right now confidently and deliberately elaborates: Schrödinger has "deeper originality and stronger creative power" than Max Born. Then he looks around waiting for any objection to stir. Nothing. Everyone is stunned. And speechless! There is a pause. Fritz Haber hesitates at first but then allows a vote and suddenly the motion is accepted, and the meeting is over. Schrödinger before Born. Max Planck laughs quietly to himself on the way home. That's how you do it! Wait patiently and shortly before the end state in a quiet voice: "there can be no doubt, dear colleagues, that… " Smooth! Born with his absurd ideas about causality, with his grotesque interpretation of Schrödinger's wave, with his crude matrix mechanics. He should stay in Göttingen for a while. Einstein himself had recognized it: Schrödinger is the genius here!

December 1926, Göttingen

Göttingen is proud. Really proud. James Franck won the Nobel Prize for Physics. For the experiment that he carried out together with Gustav Hertz and which is therefore known as the Franck-Hertz experiment. That was more than a decade ago: between 1912 and 1914, the two of them carried out their experiment, which showed very directly and beyond doubt that electrons in the atom occupy very specific, fixed energy levels. At the time, this was interpreted as a confirmation of Bohr's atomic model, the very model that was now about to be discarded and replaced by the theory of quantum mechanics. James Franck had already been awarded the Nobel Prize in November and now, in December 1926, according to tradition, hundreds of students, lecturers, and citizens marched in a torchlight procession through the winding streets of Göttingen to honor James Franck and Gustav Hertz. Today, returning football teams are celebrated by crowds after winning World Cups. In Göttingen at the time, it was Nobel Prize winners who were publicly celebrated. The torchlight procession finally ends on the market square in front of the medieval town hall. There are speeches, greetings, congratulations. Food is served, singing and drinking, and for a brief moment everyone feels as one and is proud of their Göttingen science. We are Göttingen!

And Max Born is also proud. After all, he was the one who brought James Franck to Göttingen. And James Franck was not only a good colleague to him during these years, but he was also something of a friend. James Franck and Max Born—the two were often mentioned in the same breath. The successful duo from Göttingen. But Franck was much more popular with the students than Born; he was the more amiable of the two. Always charming, always attentive, always approachable, while Born often seemed a little reserved and distant. But maybe he was just worn out. Driven by the public furor over the new quantum mechanics and surrounded by many young people who were exceptionally intelligent and ambitious, at the age of 43 he found it increasingly difficult to keep up with all the young people. It was like Goethe's poem about the sorcerer's apprentice. He suffered from the spirits he himself had summoned. Max Delbrück put it this way:

> *He suddenly found himself at the center of the maddening vortex of the quantum mechanics breakthrough, where ideas were tumbling around at a breathtaking pace.*[59]

Somehow this whirlpool robbed him of his senses and all the smart young people had frayed his nerves. It is quite possible that this was one of the reasons that Max Born turned a little more reserved than was usually his style. And so some students preferred to go to James Franck with their problems.

December 1926, Copenhagen Versus Göttingen

On December 2nd, Paul Dirac sends an article to the Royal Society from Copenhagen. The title: "The Physical Interpretation of the Quantum Dynamics." On December 18th, Pascual Jordan sends an article from Göttingen to the *Zeitschrift für Physik*: "On a new justification of quantum mechanics." Both articles could hardly be more different in language and structure and yet their content is essentially almost the same. Two authors who know nothing of each

---

[59] Greenspan (2006, p. 153).

other arrive at the same basic theory almost simultaneously, a theory that back then was called the Dirac-Jordan transformation theory. Both articles made a lasting and deep impression on insiders—from Pauli to Heisenberg, from Born to Schrödinger. Clearly works by the crème de la crème, by the chosen among the chosen, by the absolute geniuses among the normal geniuses. Here, two people of entirely equal standing had gained particularly deep insight into the nature of quantum mechanics. They found an overarching theory, studied transformation properties, and were able to present in a very abstract form something from which practically everything else follows. A theory from which the Schrödinger equation can be derived. A theory that explains why Heisenberg's matrix theory, Schrödinger's wave theory, Dirac's and Wiener-Born's representations, and why all these complicated theories are just special cases of this more general transformation theory. A theory that allows one to recognize the closer connection, the inner interrelationship between the different representations. Decades later, in an interview, Heisenberg was asked by an American what he actually thought of these two papers at the time:

First let me say that it made a very great impression on me that one now had a mathematical scheme from which one could understand why there are so many different ways of putting the mathematical scheme.[…]. Now, by means of the transformation theory, we had this complete flexibility; we could change from one scheme to the other and even to a third scheme. One could see that at the center of the problem there was a kind of abstract algebra or group theory and then you had so many different ways of expressing the thing. So in that way I was very happy about the development.[60]

---

[60] Heisenberg W., Interview with Kuhn and Heilbron (1963).

Within no time of the circulation of these two articles Heisenberg was fully aware of their significance. He wrote a letter to Pauli:

> *Today I wanted to share with you some of the thoughts that occurred to me about the clear meaning of quantum mechanics, which has been pretty much mathematically completed by Dirac-Jordan.*[61]

That's it: mathematically pretty complete. With these two articles a turning point had truly been reached. A closure. An end to the mathematical elaboration of quantum mechanics. And Dirac himself, who was usually rather distant about his own work, called this manuscript "my treasure" and was visibly proud of the work. There was a reason for that. During his studies, he had perused in great detail the work on classical mechanics by the famous William Hamilton, who had managed to successfully combine different descriptions of the same phenomenon using a "transformation theory." And now he had achieved the same thing with quantum mechanics. He, Dirac, was practically the "Hamilton of quantum mechanics." If that doesn't make you proud, what will?

Another question that presented itself was: how can it actually be that two authors who have nothing to do with each other come up with the same theory almost at the same time? This happened to Paul Dirac repeatedly and, interestingly enough, always with Pascual Jordan. His very first paper on quantum mechanics was quite similar in many ways to the first paper by Born and Jordan. And Pascual Jordan had also worked out the same results presented in Dirac's paper on Fermi-Dirac statistics beforehand

---

[61] Letter from Heisenberg to Pauli dated February 23, 1927, see Klein and Toomer (1979, p. 376).

and then Max Born traveled through half of America without sending his manuscript to the *Zeitschrift für Physik*. And now for the third time a simultaneous finish: Paul Dirac versus Pascual Jordan. How is something like this possible?

One can only speculate here. Quantum mechanics is a large collaborative effort and work is progressing in such a way that questions arise. Obvious questions that many people see immediately and less obvious questions that only a select few harbor. If two people who are similarly gifted and disposed recognize the same questions that others do not see, recognize them at the earliest possible moment and find them equally interesting and then set about solving these questions at a similar pace, then the resultant identical outcome is no surprise. That is to say: Dirac and Jordan seem to have had fairly similar abilities and talents and found the same questions similarly interesting.

But it is important to add: they both had the same person in the background who whispered an interesting question in their ear at almost the same time. This whisperer was Wolfgang Pauli, who had already made his mathematical joke in his October letter to Heisenberg, Dirac, and Hund in Copenhagen, namely the transformation of the wave function from a q-representation into a p-representation. That must have given Dirac the idea. And it was Pauli who directly influenced Jordan with this idea in a letter, as Jordan emphasizes several times in his article. Wolfgang Pauli is a great inspirer who works behind the scenes via letters and gets others going with his ideas. But perhaps one can also take a broader view: the deeper mystery surrounding the emergence of quantum mechanics probably has to do with the fact that a group of fairly similarly gifted people had coalesced, people who inspired each other and allowed themselves to be inspired, people who had similar thoughts and wanted similar things, in short: people who knew how to impact each other and thus how to have an impact together.

This publication probably marked the apex of Jordan's life achievements. After 1927, Jordan drifted away from Quantum Physics, never to regain his caliber of that time. His slide into fascism coincided with the dramatic decline of Jordan's relevance. He wasn't even particularly well accepted by the Nazis themselves. Because of his association with Quantum Physics and with so many prominent Jewish scientists, the Nazis were suspicious of him. He was semi-banished to a minor professorship in Rostock and was never really involved in any major way in the war effort. Despite the usefulness of having such an enthusiastic supporter, Jordan's enthusiasm was not returned by the regime. In this as in many other things, he was eclipsed by Heisenberg, who, even though was not a supporter of the Nazis, was much more deeply involved in the German weapons program. His post-war reputation also fared better as he kept his political views to himself, unlike Jordan, who dropped his pseudonym as soon as the Nazis came to power. Whatever he claimed after the war, Jordan proved in the 1930s that he was in full alignment with the party. In his 1941, book *Die Physik und das Geheimnis des Organischen Lebens* he wrote a glowing assessment of the Nazis, crediting them with bringing together a "ripped and divided Germany." He even joined the SA (Brownshirts) in 1933, attaining the rank of *Rottenführer* (squad leader), something that was by no means required for professors. Pascual Jordan was, it has to be said, someone who could calculate everything correctly, but had made a complete miscalculation when it came to his life.

We can say that Jordan's later life mirrored the complexities of so many others. After the war, he tried to distance himself from his earlier views but with limited success. He didn't work for 2 years and wasn't able to get a full professorship again until 1953. With the help of Pauli and Heisenberg he was able to obtain a so-called "Persilschein," a whitewashing of his earlier activities through personal recommendations and endorsements that allowed him to re-enter professional society, even to the extent of becoming a member of parliament for the conservative CDU party in 1957 under Konrad Adenauer. His political career suggests that his militaristic views had not radically changed. He wrote articles that supported the arming of Germany's military with nuclear weapons, in contradistinction to the so-called "Göttingen Eighteen," who argued against it and included many former friends and colleagues like Max Born and Heisenberg. His relationship with his former mentor was severely strained by this episode, with Born's furious wife Hedi even writing an article in the *Deutsche Volkszeitung* titled "Pascual Jordan, Propagandist on the CDU Payroll." His scientific activities never again reached the heights of the 1920s, though he mentored several scientists who would later go on to become famous themselves.[62]

December 1926, Dessau

A large, connected glass front facing the street! Unique! The Bauhaus building in Dessau, designed by Walter Gropius personally, is finished and will be ceremoniously inaugurated on December 4th. Built almost at the same time and also designed by Gropius is a housing development consisting of three semi-detached houses and a single, detached house. They are intended to serve as homes for the director and the masters of the Bauhaus, combining living and

---

[62] Schroer (2003, p. 5).

working spaces and thus also reflecting the new architectural program in other respects.

Already in the spring of '24, the new conservative State government in Thuringia had thrown down the gauntlet to the Bauhaus and brutally cut its budget by half. Such a drastic budget cut is effectively a termination. And that's exactly what was intended. But the Bauhaus, founded in Weimar, had been known long enough to be able to choose a new home. A number of other cities immediately offered to host the Bauhaus, for example, the city of Cologne under the busy mayor Konrad Adenauer. But in the end they decided on Dessau. A still young industrial city and therefore an excellent fit for its own program. A stable SPD party majority in the city council as well as the support of the famous airplane engineer Hugo Junkers who believed in promoting art and helped facilitate the move.

The Bauhaus was the Bauhaus. It was itself. Something completely new, a unique amalgamation of arts and crafts, architecture, and design. The avant-garde of modernity in all areas of applied art, with a lure that can hardly be described. Well-known artists joined the Bauhaus: Oskar Schlemmer, Lyonel Feininger, Wassily Kandinsky, Paul Klee, Marianne Brandt, Wilhelm Wagenfeld, Marcel Breuer, and Ludwig Mies van der Rohe. The absolute pinnacle of the art and design world united around a shared social ideal, an ideal that is perhaps lost today: that everyone deserves a beautiful place to live, that social housing and municipal buildings should be designed with care and attention to the people who live and work in them. Teachers and students united in noble and stirring idealism, striving to combine the useful with the beautiful. Functional and inexpensive prototypes with their own aesthetics; the birth of product design. And after moving to Dessau, collaboration with industry really began. Furniture made of tubular steel, the first cantilever chairs, and the preoccupation with mass housing construction as a solution to the social problem in growing cities. And a long-planned book series "Bauhaus books" that was finally realized in Dessau, along with the magazine "Bauhaus." Things were really gaining momentum.

December 1926, Copenhagen

Dirac had never used a telephone before and was very embarrassed when his landlady handed him the receiver. She conveys that Professor Bohr is on the line. Hesitatingly Dirac picked up the receiver and mumbled "Yes, please?" He had to speak louder, Bohr now shouted, the connection was catastrophic. Then he continued shouting and talking about the approaching Christmas holiday. Would Dirac perhaps like to spend Christmas with him and his family? Dirac was amazed and remained silent. So Bohr kept screaming. Did he understand him? He was asking whether he wanted to celebrate Christmas with him? "Yes! Thank you!" Here Dirac actually spoke two words in a row. He accepted. The invitation was fixed. So Dirac spent the holiday with the Bohrs.

His parents in Bristol, on the other hand, only had their daughter Betty at home for the celebration, a rather bleak affair. The parents' relationship was strained, his mother felt unhappy at Charles' side, and suffered from the never-ending housework and the lack of company. To get away for a few hours, she snuck out in the evenings with Betty to French class and otherwise wrote long letters to Paul in which she talked about her loneliness. And, surprisingly, his father also wrote to Paul, and specifically on black-bordered paper, signifying mourning. Apparently he had softened after Felix's death and suddenly started courting Paul, his "only son," as he wrote. Declared that all his thoughts were with him and whether Paul could tell him about his work. Might he gift Paul a chess set? And then—Paul Dirac had to read it twice because he couldn't believe his eyes—Charles Dirac signed this Christmas letter with the words: "Many kisses from your loving father" Kisses! Yes, Charles Dirac actually wrote the word "kisses." What's up with that, Paul thinks, shakes his head in surprise and puts the letter aside.

And then Christmas in the cheerful, spirited, funny, and easygoing Bohr family, with the emotionally stunted Dirac in the middle. But this experience with the Bohr family left a deep impression on him: what a nice family life! Is there perhaps a life outside of science? Bohr, for his part, overlooked Dirac's strange behavior, his spartan conversation, and his obscure black-and-white views, at least for the holidays. But years later he habitually entertained his guests with hilarious Dirac stories.

But the Bohr-Dirac time in Copenhagen was now inevitably coming to an end. At the end of January, Dirac was supposed to leave for Göttingen to complete his second post-doctoral period with Max Born. In January, Dirac quickly began to merge Maxwell's classical theory of electromagnetism with quantum theory in the sense of a higher-level theory, i.e. laid hands on what would later be called quantum field theory, in which and with which particles and fields are uniformly described. But that is a whole other chapter that follows the origin of quantum mechanics and does not belong here.

And by the time he finally put Dirac on the train to Göttingen, Bohr knew what kind of person Paul Dirac was: "probably the most remarkable scientific mind to have appeared in a long time," "a completely logical genius." And Dirac, in turn, had a very kind take on Bohr. We do not know whether Dirac already proclaimed Bohr as the "Newton of the atom" and the "most profound thinker I have ever met" back then. In retrospect, however, he did. And as we know: Dirac is not prone to exaggeration.

1927, Synthesis: Dualism and Uncertainty

January 1927, Copenhagen

The new feeling of life in modern times is characterized by an underlying unease, a basic feeling of uncertainty. Of all the works that deal with this topic, Franz Kafka's *The Castle* is possibly the most expressive. *The Castle*, published just a few months ago, is a strange, surreal novel. Depicting a system of bewildering complexity over which even its leaders no longer have control. A completely impersonal surveillance system that serves the sole purpose of maintaining its own power, what Hannah Arendt described as "tyranny without tyrants." Kafka's work gave rise to the modern term "Kafkaesque," which describes an unnecessarily complicated bureaucracy, the benefits of which are no longer clear even to the officials who run it. Kafka's characters always strive to escape the labyrinths in which they live, but the disorientation that his works describe is always a cry for a new connection to a world that seems to have lost its former, its natural, its old meaning.

A system of bewildering complexity, a labyrinth that one wants to escape, a cry for a new meaning in the world—all of this is a not inappropriate description of the situation in which the emergent quantum mechanics found itself in these weeks. The theory was actually already finished. Mathematically speaking, everything had been gathered and was now actually written down in final form; at least since the papers by Jordan and Dirac in December 1926. However, the discussions about the meaning—and interpretation—of this theory were in full swing. Pascual Jordan puts it this way in his memoirs:

> *The ice was now broken. New advances in penetrating the quantum mysteries were continually being made; But in a most peculiar and exciting way, despite the accumulating and even rapid individual successes, the problems did not fundamentally become any more transparent, but only became more and more mysterious. It became increasingly clear that an undreamt-of deeper layer of natural secrets had been discovered; that completely new thought processes that went beyond all previous physical ideas were required if one wanted to resolve the apparent contradictions that were coming to a head here.*[1]

The drama "Heisenberg versus Bohr or: the discovery of the uncertainty principle" took place in three acts: before, after, and during Niels Bohr's skiing vacation, which began on February 12th and ended on March 14th. The drama did not lack any of the typical dramatic elements: deep insights and great errors, unity and violent arguments, arguments to the point of deep exhaustion, inspiration, and quarrels, appreciation and insult. Great relief when they could finally avoid each other (because of the skiing holiday!), and later great shame because they had offended the other person so much. This drama was probably the really

---

[1] Jordan P., Begegnungen (1971, p. 18).

1927, Synthesis: Dualism and Uncertainty 185

painful last contractions, the final push giving birth to quantum mechanics.

Ever since Schrödinger left Copenhagen, the topic of "What have we actually discovered?" continued center stage. And since there was no longer a Schrödinger, sick and stuck in bed, to talk to, Niels Bohr now grabbed his assistant, Heisenberg, and harrassed him with his views and questions. Heisenberg reports in his autobiography:

> *In the following months, the physical interpretation of quantum mechanics was the central topic of discussions between Bohr and me. At that time I lived on the top floor of the institute building in a nicely furnished little attic apartment with sloping walls from which you could look down on the trees at the entrance to the Fælledparken. Bohr would often come to my room late at night and we would discuss all sorts of so-called thought experiments to see whether we really had a complete understanding of the theory.*
>
> *It soon became clear that Bohr and I were looking for a solution to the difficulties in slightly different directions. Bohr's efforts were to allow the two visual ideas, the particle image and the wave image, to stand side by side on an equal footing, whereby he tried to formulate that although these ideas were mutually exclusive, only together did both enable a complete description of atomic events.*
>
> *I wasn't comfortable with this way of thinking. I wanted to assume that quantum mechanics in its then known form already prescribed a clear physical interpretation for some of the quantities that occurred in it and that in all probability one no longer had any freedom with regard to the physical interpretation. Rather, one would have to be able to determine the correct general interpretation through clean logical reasoning from the more specific interpretation that is already available.*[2]

---

[2] Heisenberg W, Der Teil und das Ganze (2022, p. 95).

This is all phrased very politely and cautiously. "I didn't like this way of thinking"—is a great way to put it. The translation probably means: "I couldn't do anything with it." In reality, Niels Bohr must have been incredibly annoying. He really wanted to get philosophical and began to see problems that weren't even there. He, Bohr, had to endure all these smarty pants the entire time, from Kramer, Pauli, and Heisenberg to Erwin Schrödinger and Paul Dirac. He had to witness these young whippersnappers pull off one coup after another. He himself had not yet been able to make any real contribution to the new quantum mechanics, which was no surprise, since Niels Bohr was a fairly untalented mathematician.

But now he believes that his hour has come. Finally, the wisdom of the experienced man is in demand. And so he began to philosophize, which above all means: to lecture relentlessly in order to finally score points for himself. These youngsters can now please clear the stage, now interpretation is required, now leadership is required, now philosophy is required, now wisdom and overview are required, in short: now the time has come for Professor Bohr. And Bohr throws himself into the wave-particle dualism and in these weeks develops something that has gone down in the history of physics as the "complementarity principle." Wave images and particle images should be complementary to each other.

The problem with this Bohr principle of complementarity is exactly what Heisenberg quietly suggests, namely "that these ideas are mutually exclusive." And of course they are. Bohr acknowledges the contradiction, but then puts a semantic band-aid on the logical wound and says that a complete description of the processes in the atom is still possible. Hmm, you wonder. And exactly how? So Bohr looks at you meaningfully and says solemnly: Precisely through complementarity! And again, you think to yourself: I don't understand! Even his contemporaries didn't understand it. Einstein simply ignored it. Later called it Bohr's "calming philosophy." Others remained silent. Even the mean term "brainwashing" circulated. But most people simply parroted Bohr. But the principle of complementarity was somehow a completely messed up affair. "In no way is complementarity a truly understandable concept." Carsten Held expresses this fact with these fine words, and he knows what he's talking about. His book "The Bohr-Einstein Debate" cannot be surpassed in terms of diligence. Carsten Held states:

> *Complementarity, the key concept of Bohr's interpretation, has no clear meaning from the start. [….] The term is unclear because according to Bohr's statement it should not have the characteristic of contradiction, but according to his explanations it does.*[3] *[…] To this day, complementarity is often viewed as a solution to the problem of waves and particles. But to us it appears in a completely different light, as in itself problematic and unclear.*[4]

How annoying that must have been for Heisenberg: the respectful younger man who always has to listen politely, basically listen to someone talk at you in a narcissistic and somewhat confused manner. "This way of thinking was not pleasant." That immediately makes sense. That definitely wasn't pleasant. For Heisenberg there was no problem at all with the interaction between waves and particles: for him the wave function was simply a mathematical auxiliary quantity. Intended to calculate probability distributions. A formal help. Nothing else. In his opinion, this wave had absolutely nothing to do with reality! The particle concept

---

[3] Held (1999, p. 56).
[4] Ibid. p. 60.

holds true in the atomic sphere as well as in the classical sphere and the particle simply remains a particle, nothing more. Sometimes, when he wanted to annoy Bohr a little, he would add at this point that even Bohr's correspondence principle requires that a particle remains a particle.

But before one hastily dismisses Niels Bohr as a vain idiot, it should be remembered that he was a seeker throughout his life and always acknowledged this role. You can only do justice to the likes of Niels Bohr if you keep remembering the quote from Friedrich Hund mentioned above:

Bohr wasn't afraid to talk about things that couldn't actually be talked about yet.

This sentence should have been written on Niels Bohr's tomb. That characterizes this man. Niels Bohr always had the courage to search, to speculate, to sometimes find and to sometimes be wrong, and in this search, he was completely himself. That was his trademark. He captivated his contemporaries, inspired them so much, but also occasionally annoyed them immensely. And even if Bohr talked a lot of nonsense in this case, he still greatly stimulated the young Heisenberg and thus drove him to one of his top achievements.

February 1927, Copenhagen

It's February and the drama is nearing the end of the first act. The tone becomes more tense, your nerves begin to fray, and you have to get a hold of yourself more and more. The old man gets on Heisenberg's nerves because he doesn't really have anything to contribute. And the young man's talent gets on Bohr's nerves, as his polite and respectful

1927, Synthesis: Dualism and Uncertainty

manner still makes him feel that he can't really keep up. Heisenberg in his memoirs:

Since our discussions often extended until late after midnight and did not lead to a satisfactory result despite the efforts that had continued over months, we found ourselves in a state of exhaustion which, given the different schools of thought, sometimes caused tension.[5]

Tensions, a total state of exhaustion. That was to be expected and is easy to imagine! The old man's eager wooing, the young man's annoyed listening. Sometimes too harsh a criticism here, a hurtful word there. At some point, it just can't go on anymore. And Bohr decides to retreat. He originally wanted to take Heisenberg with him on his skiing vacation. But, no, better not. On February 12th he leaves just with his family and the second act of the drama begins, the act of deep insights. Heisenberg:

Therefore, in February 1927, Bohr decided to travel to Norway for a skiing holiday, and I was also very happy to be able to think about these hopelessly difficult problems alone in Copenhagen.[6]

Heisenberg was really relieved to be able to stay alone in Copenhagen. Four weeks without incessant questioning! Bliss!

And then suddenly the breakthrough. Somehow, incessant Bohr actually had an effect on him. He realizes that his quantum mechanics needs a clear interpretation and now he finally has an idea of how to pull it off. On February 23rd, he wrote to Pauli and outlined in 14 pages exactly what he would then publish a month later: the uncertainty

---

[5] Heisenberg W., Der Teil und das Ganze (2022, p. 96).
[6] Ibid. p. 96.

principle. The paper will later be called "On the clear content of quantum theoretical kinematics and mechanics." The letter to Pauli is a kind of test run. If I can convince Pauli, I can convince anyone. "Now I hope for your unsparing criticism!" Pauli's answer has become famous and indicates that the genesis of quantum mechanics is nearing its end: "Dawn of modern times. It's daybreak in quantum theory."[7]

So how does his idea play out? First of all, Heisenberg reminds us in his letter that the concept of an electron orbit was deliberately abandoned. It was his starting point, his premise. That's how it all started, back in Heligoland. And instead of that, a quantum condition was placed at the beginning. Firm and immovable, namely what was then recognized by Born and Jordan as the famous commutation relation, i.e., the formula that is written on Max Born's tomb: momentum p times location coordinate q minus location coordinate times momentum is not zero, but something with the Planck constant and the complex number i. The entire quantum mechanics rests on this relation; it is its foundation. It is its basic law. But the word "law" is wrong somehow, because this quantum condition is not a natural law, but rather it is a man-made setting. So it is, so to speak, the "Constitution" of quantum mechanics.

And if there is no longer an orbit and instead this strange relationship that mathematically links position and momentum, then the question arises: What does the connection between p and q actually mean in terms of the orbit? What does location and speed actually mean? And since speed and momentum are actually almost the same thing, one might also ask: What do position and momentum actually mean? Heisenberg puts it this way: If the commutation relation is to hold, "we have good reason to be

---

[7] Heisenberg W., Interview of Werner Heisenberg (1963).

suspicious of the uncritical application of the words location and speed."

And this suspicion is completely justified. You can no longer determine position and speed (or position and momentum) at the same time and in addition determine them with any desired degree of precision. That notion is long gone! You can determine a location only with a limited degree of precision, a—as it is called—uncertainty. And also the speed can only be determined with limited accuracy. And the uncertainty of the two quantities is related to each other. Their product is a constant. So this is the famous uncertainty principle: the product of spatial uncertainty and momentum uncertainty is a constant. If I make one uncertainty smaller, the other one—because the product is constant—becomes correspondingly larger. Imagine a rectangle and consider the cell of the rectangle. Let one edge length represent the uncertainty of the location coordinate, the other edge length is the uncertainty of the momentum coordinate of the particle. Then the uncertainty principle says: the particle with its position and its momentum is somewhere in the cell, but it is impossible to know exactly where in the cell it is. That's just a matter of chance. All we can know is that the particle is somewhere in this cell. Our ability to determine location and speed of a particle extends down to a degree of precision given by the edges of the cell, but within the cell chance reigns. The uncertainty of the two variables does not necessarily have to be of the same size. You are at liberty to determine the location with greater precision. Then the corresponding edge length becomes very narrow, but the other edge length becomes necessarily longer because its product, i.e., the area of the cell, is constant. The more you narrow down the possibilities for one variable, the more you broaden the possibilities of the other.

Heisenberg not only discusses this relation very carefully, but above all he also shows that it is a direct consequence of the commutation relation between p and q. And since the commutation relation is the foundation on which the entire quantum mechanics rests, the uncertainty principle, as its direct child, is also a fundamental equation of the entire quantum mechanics. This means that the uncertainty principle is supported by the entire quantum mechanics. If even just one experiment refutes the uncertainty principle, then the basic assumption of quantum mechanics is refuted and thereby quantum mechanics in its entirety is refuted! As long as quantum mechanics represents a valid description of nature, the uncertainty principle reflects a fundamental inaccuracy, a natural uncertainty, so to speak. Heisenberg puts it this way:

> *All experiments that we can use to define the terms "electron location, velocity" necessarily contain the uncertainty specified by the uncertainty principle. If there were experiments that simultaneously enabled a "sharper" determination of p and q than corresponds to the uncertainty principle, quantum mechanics would be impossible.*[8]

One is tempted to believe that the uncertainty principle is just a measurement problem and that it says something about the precision of modern measurements. "The picture is out of focus, can you please just focus it!" No, we can't. It is naturally blurred. This is probably why the term "Indeterminacy relation" is more precise and appropriate than the word "uncertainty principle." What is important is that it is a fundamental limitation, the most precise thing that can be meaningfully said about an object. Beyond this limit, there are no sharp and definite values of the

---

[8] Heisenberg W., Über den anschaulichen Inhalt der quantentheoretischen Kinematik und Mechanik (1927, p. 179).

1927, Synthesis: Dualism and Uncertainty

corresponding quantities. The sizes are by their nature fuzzy and indeterminable.

In this cell, defined by the uncertainty principle, the particle is wherever it would like to be. The electron moves from one point to another within this cell without us being able to describe it, without any law that governs the process. Chance, and only chance reigns here. An effect, but not a cause. The end of causality. If you work with macroscopic-size systems, where this entire cell is perceived as very tiny, then this element of chance is completely irrelevant to you as you simply perceive the entire cell as a point and the location and speed are determined via this point. On the other hand, if you deal with an atomic scale where this same cell appears large to you, then you have to live with this chance delivered by nature. And in that case, you have to accept working with probabilities as descriptive variables. Because of this chance as part of Mother Nature, quantum theory is necessarily statistical in nature.

Let's review the whole argument in one fell swoop: quantum mechanics is essentially based on the commutation relation, which was initially set as a quantum condition. The commutation relation, on the other hand, leads directly to the uncertainty principle. The uncertainty principle establishes that chance prevails in a small cell and that chance is an essential part of the theory. In short: the commutation relation already determines the probability interpretation of the wave function and makes quantum mechanics a statistical theory. For a certain state of the atom, instead of an electron orbit, there is a probability distribution for the locations of the electron determined by the wave function. This wave function obeys a wave equation, the Schrödinger equation. This dynamic gains its significance and can be tested empirically through observations of localized particles. Just as in Heisenberg's case the edges of the cell provide the smallest determinable frame, within which indeterminable chance prevails, Schrödinger's wave function is to be understood as the frame for the dynamics of the electron. Unfortunately, the behavior of the individual electron cannot be predicted because it obeys chance. But many electrons in succession or a single electron over a long period of time ultimately reveal the pattern that is predicted by the wave function.

Since the commutation relation is set right from the beginning, one could object that perhaps a better premise could be found, a better beginning, a better theory that would banish the unpopular random chance and reduce quantum mechanics to a "statistical transition theory." Well, it could be. But first find it! As already stated: this premise as well as the uncertainty principle is supported by the entire structure of quantum mechanics. If the uncertainty principle collapses, the entire QM collapses. And that is hardly conceivable anymore given the spectacular success of this theory, with its thousands of successfully explained experiments. It is extremely unlikely that a better theory will be found. It is much more likely that the uncertainty principle really reflects nature, and that chance actually prevails in the cell. In any case, Heisenberg himself viewed his uncertainty principle from the beginning as proof that the new quantum theory is a fundamental theory in such a way that it represents a complete description of atomic phenomena.

March 1927, Copenhagen

On March 14th, Niels Bohr returns from his skiing vacation and the third act of the drama begins. Let's let Heisenberg report for himself:

1927, Synthesis: Dualism and Uncertainty

There were once again difficult discussions when Niels Bohr returned from his skiing holiday in Norway. Bohr had continued to explore his own ideas while away and, as in our conversations, tried to make the dualism between the wave image and the particle image the basis of the interpretation. At the center of his considerations was the concept of complementarity, newly coined by him, which was intended to describe a situation in which we can understand one and the same event from two different perspectives. Although these two ways of looking at things are mutually exclusive, they also complement each other, and it is only through the juxtaposition of the two contradicting ways of looking at things that the descriptive content of the phenomenon is fully exploited.[9]

"The two perspectives are mutually exclusive, but they also complement each other." Two things that are mutually exclusive cannot complement each other. You can sense between the lines how much poor Heisenberg contorted himself to somehow present this obvious nonsense in an appreciative way. When Bohr came back from vacation, Heisenberg's manuscript on uncertainty had not yet been sent off but was almost finished. Bohr must have read it immediately and must have recognized that Heisenberg had simply done it better than he. But he didn't want to admit that, even just to himself. In his book, Tobias Hürter describes what happened next:

Bohr deems his idea to be the better one and Heisenberg's uncertainty principle just a special case. Heisenberg is tired of disagreeing with him. For a few days he manages to avoid further discussions with Bohr. Bohr hopes that Heisenberg realizes on his own that he has gone astray. But Heisenberg remains steadfast. Bohr urges him to withdraw his article. Heisenberg

---

[9] Heisenberg W., Der Teil und das Ganze (2022, p. 98).

bursts into tears. It hurts when a son breaks away from his father.[10]

He bursts into tears? Hard to believe; did it really happen? The whole affair does not paint a flattering picture of the Great Dane, who in turn must have completely gone astray. To recommend that Heisenberg retract his article, which later became so important in the history of physics. How embarrassing for Bohr! And he must really have meant it and used all the force of his personality to put pressure on Heisenberg; the atmosphere must have been terribly tense and the tone rude. But Heisenberg remained steadfast. Maybe his tears were also tears of anger: retracting an article with such deep and clear insight because the boss himself wants to publish something completely incomprehensible. That is the reason for anger! As a compromise, he finally asks Heisenberg to add an addendum to his article. Well, to say what? Anyone can read this addendum today: two petty points are mentioned, thanks are obediently expressed again, and it is vaguely stated that Bohr gained insights that allow for a significant deepening and refinement of the analysis presented here. In short: Heisenberg kowtows to Bohr and yet says absolutely nothing. This paragraph is somewhat embarrassing for Heisenberg and implicitly a devastating criticism of Bohr, because the emptiness of his objections is so apparent.

But Heisenberg, on the other hand, may not have been an easy partner either and may have provoked Niels Bohr in his own way. On May 16th, he wrote to his parents that there had been "serious personal differences" between him and Bohr and on May 31st he wrote a letter to Wolfgang Pauli:

---

[10] Hürter (2021, p. 227).

I want to write about what has moved me more in the last few weeks than anything else in a long time. I found myself in a fight for the matrices and against the waves; In the heat of this fight, I often criticized Bohr's objections to my work too harshly and, unknowingly and without wanting to, hurt him personally. When I now reflect on these discussions back then, I can very well understand that Bohr was upset about them.[11]

After he had already left Copenhagen in June 1927, Heisenberg then wrote a letter to Bohr in which he confessed his shame at having "appeared so ungrateful to him."

I still think about how it all happened almost every day and I'm ashamed that it couldn't have been any other way.[12]

Niels Bohr is a figure who cannot be conclusively explained. The big driver of quantum theory, the one who brings in young, highly gifted people and puts them on the scent, the man with the right nose, the inspirer and coordinator of the new theory. In many ways he is the counterpart to Max Born in Göttingen. But unlike Bohr, Born actually made his own contributions to quantum mechanics and was able to keep up as a mathematician. And unlike Bohr, he didn't play the role of the great chief interpreter but instead exercised gracious restraint. Bohr's final role as the great explainer of a theory that he had suggested but hardly helped to develop was and remains somewhat questionable.

---

[11] Letter from Heisenberg to Pauli dated May 31, 1927, see Klein and Toomer (1979, p. 396).

[12] Hürter (2021, p. 228).

April 1927, Göttingen

Quantum theory belongs to the Planckstraße, said Born, knowing full well that the namesake of his street was not Max Planck. Born had rented a ground floor apartment in the Planckstraße near the institute that consisted of three "huge rooms and an equally huge kitchen." The largest room was a living room, study, and music room in one. Additionally, a "veranda with stairs leading to a park-like garden with old trees and large lawns." Max Born occasionally held his advanced seminar on quantum mechanics in the large living room and study of his stately apartment on Planckstraße. The rooms of his beautiful apartment always radiated a peacefully warm atmosphere that immediately put everyone in the right mood to tackle this difficult theory. In other words, the cool Born of quantum wisdom originated in this warm living room.

The first and last gatherings of the semester were always special events. The Borns invited people to a small luncheon buffet and not only the students of the senior seminar attended, but actually everyone who felt they belonged to Max Born. And there were many. This was also the case on April 20th, the official start of the summer semester of 1927. Hedwig and Max Born had invited many people and everyone showed up. Four large tables were available, two in the dining room and two on the veranda, and those who had already served themselves at the buffet sat down at one of the tables. Four tables because there were four groups in Göttingen. The guests at table one were mostly associated with Prof. Pohl and called themselves "polished" (as in "Pohl-ished"), those at table two were in a similar play on words "borniert"(narrow-minded)" due to their work with Max Born. The Courant employees called themselves "courageous" and the final table seated "frank" guests, who identified with James Franck. Young people in Göttingen were either narrow-minded, frank, polished, or courageous.[13] And very proud of it! And the four associated professors were the so-called "party bosses" (as in political party) who, fortunately, liked to party a lot and always invited their students to their parties. The word "party boss" is generally a derogatory one, but in association with these professors it was probably meant affectionately. Pascual Jordan makes it clear that:

Göttingen in those days was characterized by a rare kind of human generosity: by the complete absence of 'boss cult' and 'boss attitude', as it was often drastically called.[14]

Göttingen, however, was occasionally also like an army camp, a combat force. And people went to war mostly against Schrödinger! On a walk, Born read to his disciples a letter from Schrödinger. Jordan remembers:

Oppenheimer later told me jokingly that Born had summoned his troops like a general and then explained what was wrong with Schrödinger's ideas.[15]

Oh Göttingen, how beloved you were back then!

In the midst of the dark catastrophes of the First World War and of Hitler's Reich, an intellectual life flourished here, which rose to unbeatable intensity in light of the uniqueness of an enormous scientific task.[16]

---

[13] These are puns on German words based on the surnames: *Poliert* for Pohl, *Borniert* for Born, *Couragiert* for Courant and *Franckiert* for Franck.
[14] Jordan P., Begegnungen (1971, p. 19).
[15] Jordan E. P. (1963, p. 23).
[16] Jordan P., Begegnungen (1971, p. 14).

That's how Jordan presented it. Göttingen, in those years, was without question the European Mecca of theoretical physics and the heart of it was Max Born's apartment with its wonderful garden. The obvious meeting point for all those who were interested in quantum theory. What Niels Bohr was in Copenhagen, Max Born was in Göttingen. Anyone who wanted to pursue an academic career, who wanted to belong, who wanted to be seen, who wanted to be counted among the international physics elite, absolutely had to show up at one of Max Born's parties. Thank God, such an invitation was not only an honor but also funny and entertaining as there were always many stimulating conversations, almost always accompanied by a small musical performance. Not one, but two Steinway grand pianos adorned the huge living room and if not Heisenberg, then there was always someone else proving their piano skills.

Today Max Born himself took to the keys, performed something rather moving, and ended with tears in his eyes. And once the senior seminar was over, the assembly really got jubilant and decided to go hiking. The guests were prepared for this and had brought sturdy shoes. And so they set off over one of the many hills around Göttingen. Paul Dirac must have liked it. He told his mother in a letter that he clearly preferred Göttingen over Copenhagen: there were far more opportunities for walks there. He likened Göttingen to Cambridge, just with more hills, which he liked.

Hiking over these hills they eventually reached the Nikolausberg to eat a bite together in the country inn as evening approached. And then the beer started to flow and they got into a boisterous mood, challenging each other to little competitions. The students' favorite game was clear from the start: the potato race, a race in a large circle around the tables pushed together, in which a potato was not allowed to fall from the spoon carried in front of them. And, of course, the highlight of the evening: when the four "party bosses" had a turn at this race. It was hilarious, especially because Pohl laughed so hard that he could not make the turn and fell into the bar counter.

But today, April 20th, it was agreed upon that following the professors, the assistants and postdocs should also have a potato race. Poor Paul Dirac was absolutely horrified by the idea that he would embarrass himself and he could be seen secretly disappearing down the long corridor by the restrooms with a spoon and potatoes in order to practice! With his long, lanky legs, he must have looked like an employee of Monty Python's "Ministry of Silly Walks" who keeps dropping his potatoes but returns to the task again and again with great earnestness.

Later he hikes back to Göttingen with Robert Oppenheimer, back to the same apartment because they were lodgers with the same family, and became close in those months. Oppenheimer, who secretly observed Dirac's silly walks, now makes fun of him with a biting mockery, which Dirac endures with stoic calm. That was also his reaction when Pauli made fun of him. Silence and a completely blank expression. But while he can't stand Pauli, he thinks Oppenheimer is quite ok. Albeit, he wonders about Oppenheimer's poetic streak: how can anyone this intelligent waste time reading or worse yet writing poetry? Dirac simply could not wrap his head around this. And now, as Oppenheimer finally has laid to rest the potato race topic and as they reach the outskirts of Göttingen, he asks the question he always wanted to ask:

I can't imagine how you can work in physics and write poetry at the same time. In science you want to express something that no one knew before, and in words that everyone can understand.

> *In a poem you are forced to say something with words that everyone already knows, but no one can understand.*[17]

Oppenheimer, whose answer is not known but who was known for his completely incomprehensible writing, probably replied that in both areas one had to be careful to write something that no one knew before and no one could immediately understand. That would be the desired connection between the two activities. Oppenheimer winked at him: "And when it comes to 'incomprehensibility,' you probably know what I'm talking about, don't you, Paul?" In fact, that's exactly why Oppenheimer found Paul Dirac so interesting: he wasn't easy to understand and didn't attach any importance at all to being understood. When Dirac spoke, people usually just sat there and listened to him with their mouths open. Oppenheimer was simply very impressed by Paul Dirac and later said in retrospect:

> *The most exciting time of my life was perhaps when Dirac arrived and showed me the proofs of his work on the quantum theory of radiation.*[18]

This referred to Dirac's quantum field theory, which he had brought with him from Copenhagen to Göttingen and which Oppenheimer found "extraordinarily beautiful." Robert Oppenheimer also saw a future for his own academic work in it.

But of course, everything would turn out completely differently. The name Oppenheimer today stands for the construction of the first atomic bomb. The very beginnings of development are always the most exciting. Like the smallest roots that in time, as a story progresses, become larger and

---

[17] Farmelo (2016, p. 123).
[18] Ibid. p. 126.

unite with other roots to become the cause of an ultimately great event. The atomic bombs on Nagasaki and Hiroshima are not only the story of Oppenheimer and his physicists or of an American government that resorts to extreme measures, but also of Japan, which unnecessarily enters into a second world war. And while Oppenheimer delves deeper into physics in Göttingen, the young Emperor Hirohito is working on the rise of Japan as a military and colonial power. He had been crowned in December 1926 and immediately began building up the Japanese military. All this culminated in war against China in 1931 and the atomic bombings in 1945. When Hirohito spoke on the radio for the first time in August 1945, he had only one thing to tell his people: unconditional surrender. But in April 1927, all of this was still quite far away: Hirohito was just 26 years old, Oppenheimer 23. Young men in the spring of their lives.

April 1927, Hamburg

Quantum mechanics had three parallel strands of development. The first two strands were Göttingen matrix mechanics and Schrödinger's wave mechanics, which were finally combined into a coherent overall concept in 1926 and rounded off with the uncertainty principle. But the third strand—that was the discovery of electron spin; a diagonal move across the other two. This discovery arose from the investigations into the anomalous Zeeman effect, and was linked to the names Pauli, Goudsmit, and Uhlenbeck, taking place parallel to the elaboration of the first two strands. Many had hoped that at some point spin would tumble out of the quantum mechanical theory on its own. But that didn't happen. And now people asked themselves: Where on earth does this spin actually come from? And now that all corners of the theory had been scrupulously explored, everyone scratched their heads. Had they overlooked something? How on earth are you supposed to connect the third strand to the first two? However, one last hope existed. Yes, it existed! Because up to this point, it had not been possible to combine quantum mechanics with Einstein's theory of relativity. Maybe this union would bring forth the spin.

But so far no one had been found who could marry the theory of relativity and quantum theory. It was still simply too much for the boys, and so they accepted spin as one accepts rain and snow: it was just a natural event! That's just how it is! A shrug! And even if one shouldn't imagine spin as its own rotation, the electron still has its own angular momentum and its own magnetic momentum. And again: a shrug! And this spin has two possible orientations in space and leads to the electron's own degree of freedom, which can take on two possible values. Spin +1 and Spin -1. A third time: a shrug! That's just how it is. You just have to take life as life is!

Wolfgang Pauli would have loved to be the one who brought quantum and relativity theory together and magically pulled spin out of a hat in the process. After all, it was he who—at just 21 years of age—summarized the theory of relativity in the "Encyclopedia of Mathematical Sciences" so elegantly that Einstein was amazed. A mature piece of work, he had said. And after all, it was also he who, as a midwife, initiated the birth of spin. Was it so surprising that he now dreamed of the role of fundamentally deriving spin from the combination of quantum and relativity theory? No, it wasn't. But he simply couldn't do it. And then he slowly and reluctantly came to terms with the idea and, like most others, simply accepted spin as an additional natural event and devoted himself to developing a non-relativistic spin theory. That means: This theory simply assumes spin, lets it exist from the start and squeezes, nay forces it subsequently into quantum mechanics. Not very elegant! Especially for someone of Pauli's caliber. He had already finished a first version of the manuscript "On the Quantum Mechanics of the Magnetic Electron" in November 1926. But when Heisenberg wrote to him:

I think your calculation about the spinning electron is very pretty: you should publish it somewhere![19]

He became outright nauseated. "Very pretty!" Gee! Outrageous! Now his beautiful thought turns into just a measly number. In addition, it is somehow a rehash, because Jordan and Heisenberg had already incorporated spin into the matrix theory in this superficial way a year ago and thus explained the anomalous Zeeman effect. So why should he, the great Pauli, tackle it again? On March 12th he writes once more to Jordan, thanks him for a nice suggestion that had returned to him the joy of a non-relativistic spin theory, and then and for the time being goes on a long vacation to southern Germany. But then, in April, he approaches the problem anew, takes the old paper out of the drawer and sits down at his desk.

The idea: he assumes not just *one* Schrödinger wave function for each electron, but two such functions. One function for the electron in the spin-up state and one function for the electron in the spin-down state. A two-dimensional vector, so to speak, whose components are wave functions. And the spin operators are two times two matrices, the so-called Pauli matrices. And this juxtaposition of two functions is now carried through the entire theory so that the Schrödinger equation becomes the Pauli equation, which

---

[19] Letter from Heisenberg to Pauli dated November 21, 1926, see Klein and Toomer (1979, p. 357).

technically contains nothing new at all, except that you have everything twice. The paper was submitted on May 3rd and is now seen as an important contribution to the origins of quantum mechanics: spin is forcibly incorporated into the theory. That's what the name Pauli stands for.

You can already guess who was destined to bring together quantum theory and relativity theory, can't you? Of course: it was Paul Dirac. On January 1, 1928, just nine months later, he was to submit the paper with the derivation of an equation that is now known as the Dirac equation. So there are three equations: The Schrödinger equation, if you don't care about the theory of relativity or spin, the Pauli equation, if you don't care about the theory of relativity but do care about spin, and finally, towering over everything: The Dirac equation as a fully relativistic theory. And exactly as one had already suspected: The Dirac equation allows the concept of spin to emerge as if by itself, with all the accessories that had previously been measured by the experimenters. Spin is not assumed in any way but arises from the theory. If spin had not yet been discovered at this point in time, it could have been predicted to exist theoretically based on the Dirac equation. Oh, what an equation this is! An equation, says Frank Wilczek, that is "painfully beautiful." An equation that allows you to foresee antimatter. An equation that really describes the behavior of the electron conclusively, the global formula of the electron, so to speak. And when you look at the memorial plaque for Paul Dirac in Westminster Abbey today, what do you see? His equation of course! The Dirac equation. His greatest achievement!

May 1927, Göttingen

Paul Ehrenfest, by the way, is the tragic figure in this entire story. Throughout his life, Ehrenfest had suffered from dramatically low self-esteem, which was in stark contrast to the high regard in which he was held by almost all of his colleagues. Paul Ehrenfest simply couldn't handle quantum theory, which probably cost him his life in 1933. He was already prone to depression and was also crestfallen by his son Wassik's Down syndrome. In September 1933 he shot Wassik and then himself. His last letter was addressed to Bohr and Einstein, among others, but was never sent:

> *I have absolutely no idea how I'm going to carry the unbearable burden of my life over the next few months. I can no longer bear to let my professorship in Leiden go down the drain. […] In recent years, it has become increasingly difficult for me to follow developments in physics with understanding. After many attempts, which increasingly exasperated me and tore me apart, I finally gave up in despair. That made me completely tired of life.*[20]

Futile attempts to understand the new physics had made him tired of life. After his suicide, Einstein wrote that Ehrenfest had felt no longer up to the task of research. And further:

> *This situation has been made worse in recent years by the peculiar turbulent development that theoretical physics has recently experienced. Learning and teaching what one does not fully affirm inwardly is a difficult matter in itself, doubly difficult for a fanatically honest mind to whom clarity means everything… I don't know how many of the readers of these lines can*

---

[20] van Delft (2014).

understand such tragedy; But it was primarily [this tragedy] that caused his escape from life.[21]

Is Einstein trying to say that Ehrenfest killed himself because, as a professor of theoretical physics, he found himself completely overwhelmed by quantum mechanics? Hard to believe.

In any case, in May 1927, Paul Ehrenfest was still alive and, it seemed, in the midst of the developments of quantum mechanics. All the great minds wanted to spend some time with him. Heisenberg, Dirac, and now also Robert Oppenheimer, who on May 11th will defend his dissertation, which earned him summa cum laude from Born. In addition to Born, his second examiner is James Franck, who finally received his well-deserved Nobel Prize 6 months earlier. Franck questions Oppenheimer for only about 20 minutes, and even that is far too long for him. At the end he says, visibly relieved: "I'm glad it's over. He started questioning me!" Immediately after the doctoral examination, Oppenheimer proceeds to Leiden to do his postdoc with Paul Ehrenfest. In a letter to Ehrenfest, Born finally reveals his true opinion of Oppenheimer. He was "almost mentally ruined by this man." And further:

I would like to tell you that I have never suffered under a person as I have under this one. He is undoubtedly very talented, but completely lacking in mental discipline. Although he is modest on the outside, he is incredibly arrogant on the inside. He completely paralyzed us all for three-quarters of a year with his way of knowing everything better and expounding on every thought presented to him. I am breathing a sigh of relief since he left and am gathering the courage to work again. My young people feel the same way. Don't agree to keep him for any extended period

---

[21] Einstein A., Aus meinen späten Jahren (1979, pp. 205–206).

of time. Wait, you should tell me your own opinion. Maybe I just got nervous.[22]

June 1927, Göttingen

He gets more and more nervous these days for other reasons as well. Sometime in the summer of 1927, Max Born hears the term "German physics" for the first time. A group of German physicists who reject the theories of relativity and quantum mechanics as too mathematical gather under this name. There is a contrast—according to the representatives of German physics—between an overly abstract, Jewish physics, and a comprehensible German physics. Science is not international, says German Physics, not at all. "In reality, science, like everything that humans create, is conditioned upon race and blood." This is what Philipp Lenard writes in the foreword to his four-volume textbook "German Physics." Philipp Lenard, who already dismissed the theory of relativity as a pure hypothesis in his book on the ether theory and early on argued against "Jewish influences" in physics and, in particular, against Albert Einstein. At his side is a certain Johannes Stark, whose Stark effect played a significant role in the discovery of quantum mechanics.

Scientifically, this group is not taken seriously. But it forebodes what is to come. It creates an atmosphere that sends shivers down Max Born's spine. Nascent nationalism is in the air and many Jewish researchers at his institute in 1927 have already smelled it. Is Germany in danger of slipping? Max Born is completely flabbergasted: how can quantum mechanics in particular be misused for national and racist purposes, quantum mechanics, which was created and advanced only thanks to people working together on a

---

[22] Greenspan (2006, p. 156).

truly international level, across borders and cultures? What will happen to quantum mechanics, what will happen to the Jewish scholars in Germany if this new spirit, this terrible evil spirit gains ground in Germany?

This is also the summer that Max Born presumably noticed that something was amiss with his protégé, Pascual Jordan. It is difficult to pinpoint the exact moment. It is likely, however, that Jordan only openly acknowledged his National Socialist stance in 1928, when he became a private lecturer in Hamburg, where many prominent figures belonged to the right-wing extremist scene. But given that Paul Ehrenfest somehow got wind of Jordan's beliefs in 1927 and said that Max Born had to be warned, one can assume that the news must have somehow reached Born in 1927. In any case, by the end of the 1920s, Jordan's right-wing extremist views were an open secret among physicists. And Jordan's involvement in the Nazi movement was by no means superficial. In 1930, under the pseudonym "Ernst Domeier," he published articles about culture and politics in magazines of the völkisch movement, for example, in the magazine *Deutsches Volkstum*. It was well known that Jordan was using the name Ernst Domeier, and when Hitler finally came to power, Jordan had become well known in right-wing extremist circles. The topics of his articles varied widely, but they all had a common denominator: elitist condescension towards the lower classes and a deep distrust of any kind of cultural modernity. It is also clear that Jordan fully supported the anti-Semitic views of those around him. But Jordan didn't get involved in "German physics." On the contrary, he promoted quantum mechanics among the Nazis, which he believed was a sure bulwark against what he called "the materialism of the Bolsheviks."

In 1927, some students in Göttingen brazenly wore Nazi armbands and threatened Jewish students and faculty. After the Nazis came to power in 1933, the purge was swift and merciless. Nazi student groups had already drawn up lists of Jewish professors who should be removed. Both Born and Courant immediately lost their positions and Franck resigned in outrage. The heads of the experimental and theoretical physics departments in Göttingen disappeared overnight. The extraordinary collaboration between these two departments had made Göttingen famous, but now it was over. The heart was ripped out of Göttingen as a powerhouse of physics and Born was devastated. Everything he had worked on over the past decade was in ruins. And how did Jordan see this? It is difficult to say since he destroyed many documents after 1945. But he could not have been ignorant to the fact that the Nazi ideals he had supported for so many years now had consequences. He visited both Born and Franck and complained that he did not have enough influence in the party to prevent all this. A pathetic reaction from someone who owed so much to both of them. Whatever he claimed after the war, Jordan proved in the 1930s that he fully embraced the Nazi party.

The Jewish professor Max Born, who was one of the first to lose his professorship in 1933, and his National Socialist assistant Pascual Jordan. What a pair! What might Max Born have thought and felt when he realized that Jordan was on the way to becoming a solid Nazi? He, Born, who had enabled Jordan to continue his studies. He, Born, whose compassion had led him to raise money so that Jordan could receive medical treatment for his stutter. And, conversely, what was going on inside Jordan's head when he endorsed the anti-Semitism of his Nazi party while at the same time cultivating close personal contact with personalities like James Franck, Richard Courant, and Max Born? When you look at Jordan, you can't help but be amazed. This man must have been schizophrenic: a staunch anti-Semite in an almost loving relationship with his surrogate father Max Born. In 1970 Pascual Jordan wrote an obituary for Max Born:

> *I would also like to say that, next to my parents, he was the person who had the deepest, most lasting influence on my life.*[23]

How strange that a person can pick up his convictions from so many different sources but cannot derive them from his own personal experiences in daily interactions with this obviously unique and incredible person.

It is illuminating to place the protagonists of quantum mechanics, Wolfgang Pauli, Werner Heisenberg, Max Born, Paul Dirac, Pascual Jordan, Erwin Schrödinger, and Niels Bohr next to each other. Five of these seven personalities came from professorial families, so they had a fairly similar family and middle-class backgrounds. They were all incredibly ambitious and six of them even won a Nobel Prize. They were of one mind, at least as far as physics was concerned. But beyond that, their life plans, their personal beliefs, their political attitudes, basically the most essential aspects of human life, diverged tremendously. So much so that even a Nazi could be found among these seven, who incidentally probably paid for his beliefs with a Nobel Prize. In the end, Pascual Jordan was the only one in this group who didn't get one.

Paul Dirac's time in Göttingen ended at the beginning of June. Despite his disinterest in politics, Dirac must have noticed the new Nazi presence in Göttingen. But the reason for his return to Cambridge was unrelated. St. John's College had contacted him and invited him to apply for the position of Fellow. In addition to bolstering his reputation, this would mean a small income along with room and

---

[23] Jordan P., Begegnungen (1971, p. 43).

board at college. A permanent position at the Faculty of Mathematics would probably be the next step. Dirac would then be professionally and financially secure for the rest of his life. And so: off to Cambridge! In early summer on his way to England, Dirac made a brief stop at the house of physicist and friend Paul Ehrenfest in Leiden, before finally returning home to 9 Julius Road in Bristol in mid-July. Before returning to Cambridge, he wanted to spend a little time with his parents over the summer. That meant he would sit upstairs in his room while his parents bickered and argued downstairs.

August 1927, New Jersey

The editors of *Physical Review* received a manuscript from Clinton Joseph Davisson and Leslie Germer, which will result in a Nobel Prize. Since his return from Oxford in the summer of 1926, Clinton Davisson has relentlessly pursued his internal "program of thorough search." "Thorough search"—that was exactly his thing! The entire remainder of 1926 was initially spent optimizing the process in order to bring the nickel sample from a polycrystalline to an almost single-crystalline state. Germer and Davisson then found that the electrons had to enter at the shallowest possible angle for the effect to be particularly noticeable. And then fiddle with the energy of the electrons! Low electron energy equals low electron momentum equals large de Broglie wavelength of the matter wave. And you keep trying until the wavelength of the electrons really matches the diffraction grating formed by the metal surface perfectly.

And, lo and behold, in the end there was an almost perfect match. A wavelength that corresponds to the prediction of the de Broglie relationship to within 1%. Bull's eye!

So, in the summer of 1927 in Room 7B at Bell Labs, Davisson and Germer finally provided experimental proof of the matter wave. A memorial plaque commemorates this event today:

> *At this site, the original location of Bell Telephone Laboratories, C. J. Davisson and L. H. Germer in 1927 performed the first direct demonstration of the wave-like behavior of elementary particles, predicted by L. de Broglie in 1923. The Davisson-Germer experiment provided crucial empirical evidence for the validity of the then rapidly evolving theory of quantum mechanics.*[24]

So electrons are actually waves too. De Broglie was right. America's only contribution to the emerging field of quantum mechanics thus came from Bell Labs. "For the experimental discovery of the diffraction of electrons in crystals," was the official reason, Davisson received the Nobel Prize in 1937, which he had to share with the British physicist George Thomson, who had discovered the same phenomenon at the same time using a different experimental setup. Incidentally, the experimental setup and methodology of Germer and Davisson still live on today and is known by the acronym LEED, low-energy electron diffraction, a standard method in surface physics for studying atoms on surfaces.

Normally, the physical question comes first, after which the brilliant experimental physicist thinks about an appropriate method and apparatus for a measurement to validate the effect. With Davisson, however, it was exactly the other way around: first, there was the measuring apparatus of hot cathodes, electron guns, and high vacuum tubes, which was derived from technical questions, and only then did the

---

[24] American Physical Society plaque on the west side of the Westbeth door at Bell Labs, see: Wikipedia entry for *Davisson-Germer Experiment*.

physical problem emerge: the matter wave, which fit the apparatus like a glove. And this exciting physics came into Davisson's life through two coincidences. What would have happened if 1925 had been different for Davisson, if he had not come across Elsasser's article, if there had not been an accident in his laboratory? Well, he would certainly have missed his big chance, the determined Thomson would have received the Nobel Prize alone, and he, Davisson, would have continued with his well-behaved measurements on the polycrystalline samples, would have written his boring papers and would have filled his consulting position at Bell Labs. Nobody would remember his name today. Sometimes you just need luck in life, a fate that smiles at you for a moment!

September 1927, Berlin

Berlin and its scholars: Albert Einstein, Fritz Haber, Otto Hahn, Lise Meitner, Max von Laue, Walther Nernst. And towering over all of them is the master of physics: Max Planck, who is now retiring. Who will be his successor? Who will be the intellectual leader of German physicists, who will be Germany's most respected professor? As expected, Arnold Sommerfeld, the first person on the list, declines and so the second person on the list, Erwin Schrödinger, becomes Max Planck's heir to the throne. Zurich had offered a lot to keep the famous Schrödinger in an effort to make itself even more attractive than Berlin, but in vain. In the end, Schrödinger decided on Berlin. The real deciding factor was probably a remark by Max Planck, which you can read in verse form in the Planck family's guest book because Schrödinger graced it with a poem:

Thus I often thought, and when I came anyway,

For fame it was not, I'll swear to it.
And—I only say it with a hint of shame—
Even gold alone could not beguile me.
Decisive was a single word—from long lines
Of letters, of conversations, colorful and tangled,
Revered lips spoke it,
Not beseeching though. Very brief: It would please me.[25]

In Zurich he was only Einstein's successor in his professorship. Now he was Planck's successor and at the same time Einstein's immediate colleague. Definitely a step up. And what a step: just 2 years ago he was an absolute nobody! And now, since the spring of 1926, he was world famous. Praise and thanks be to you, you beautiful quantum mechanics! You sent me your equation in the name of glory! Heavenly messengers are heralding the news.

And so in the late summer of 1927, Schrödinger moved from Zurich to Berlin and took up his position as full professor of theoretical physics at the Friedrich Wilhelm University at the beginning of the winter semester. He did not just move; he arrived. And indeed settled fairly quickly, even though Austria and Prussia were quite dissimilar. It helped that Lise Meitner was there, also a transplant from Vienna, who quickly made him feel at home in the Prussian-German metropolis. Despite all the happiness about his achievements, he also enjoyed the fact that there were so many old and great names of scholars around him and that he was able to draft a little in their slipstream.

In Berlin you only felt a fraction of personal responsibility and could disappear among the number of those who towered over you in age and reputation. And so these years were scientifically very enjoyable and very carefree.[26]

---

[25] Moore (1992, p. 237).
[26] Hoffmann (1984, p. 52).

Schrödinger was often a guest at the small concerts the Planck family held at their house. Or he enjoyed visiting Einstein at his vacation home in Caputh. Just get out of the hustle and bustle of the big city, enjoy the beauty of the Brandenburg landscape and spend hours sailing around Lake Schwielowsee, while having long, undisturbed debates with Albert Einstein. And when he was the host himself, in his house in Grunewald, Vienna sausages were served. That's right, Erwin Schrödinger's infamous "Wiener-Wurstel Evenings," and of course with no shortage of guests here either. People liked to party in the 1920s. Everyone joined in. Berlin became his second home, he later wrote, a home that he "loved because of freedom of conscience and full freedom of expression." It didn't last long, because a love based on such values couldn't survive 1933. As early as the summer of 1933, Schrödinger, not personally persecuted politically or racially, turned his back on Berlin of his own free will and took a position in Oxford. He had spent only 6 years in Berlin.

September 1927, Como

The Partito Nazionale Fascista gained 100,000 new members in just one year and the total number of its members exceeded the million mark in 1927. There are no longer any elections within the party and Benito Mussolini has thus reached the pinnacle of his power. No one is going to challenge him anytime soon. Now he can leisurely build up what he and his fascist movement want: a strong, defensive, nationalistic Italy. However, this requires national heroes, great Italians, that the common people can identify with and who make the supremacy of the nation truly tangible. Alessandro Volta, whose 100th anniversary of death is being celebrated with a "Volta Century Exhibition," fits the bill. The Italian physicist was the inventor of the voltaic column, i.e., the electric battery, and one of the founders of the theory of electricity and, for the purpose of immortalizing his fame, had to relinquish his last name to the physical unit for, as well as the concept of, voltage. For Benito Mussolini, Volta was the ideal candidate for a new national hero. Lived and died in Como, and thus Como had to make an appearance as part of the program for the "Volta Century Exhibition." What could be more natural than inviting the great physicists of the present to Como to honor a great physicist of the past? So that they may advance physics right there, as Alessandro Volta had done before them.

In short: in mid-September, physicists from 14 countries, including 12 current or future Nobel Prize winners, gather for an international conference in honor of Alessandro Volta at the Istituto Giosuè Carducci in Como. From Heisenberg to Born, from Fermi to de Broglie, from Planck to Sommerfeld. All are there, with only one notable absence: Albert Einstein, who doesn't like Benito Mussolini adorning himself with his name. This is a bitter disappointment for Niels Bohr, because in Como he planned to make it happen. Just as he had done in the summer of 1922 at the famous Göttingen "Bohr Festival," where the worldly audience reverently listened to his seven lectures explaining the theory of atomic structure. That's exactly the role he wants to play once more in Como: the quantum general overseeing the battlefields of the new theory. Bohr, the "Newton of the Atom," the chief interpreter, the philosopher, the great mind, the man who truly sees through and behind things. At least that is how Niels Bohr envisions his part in Como. He wants to present an all-encompassing summary of the new quantum mechanics, including the final and ultimate interpretation of the formalism, which, he hopes, will only become truly rounded and beautiful through his own ideas. And now his great counterpart is missing: Albert Einstein didn't show up.

Since Heisenberg presented his uncertainty paper in March, Niels Bohr has worked doggedly on his ideas on the principle of complementarity. But he has to revise the speech manuscript for Como over and over again; he just can't get it right. And when he finally delivered his speech on September 16th in the large ballroom of the Istituto Giosuè Carducci, despite all this preparation, the lecture was "difficult," to put it diplomatically, or "incomprehensible," to put it bluntly. Nevertheless, Max Born and Werner Heisenberg stood up immediately at the end of the lecture and pronounced that they completely agreed with Bohr's view of quantum mechanics. Everyone is impressed, yet no one really understands the whole thing. But there is also simply no one in the audience willing to object. Objections arise only a month later, once Einstein participates again. In any case, this lecture in Como created what would later enter the literature as the "Copenhagen interpretation of quantum mechanics."

"Copenhagen interpretation" because the Copenhagen faction, essentially Bohr, Pauli, and Heisenberg, are considered the originators of this interpretation. This is a bit unfair, however, in that Max Born gets short shrift. After all, it was Max Born who in the summer of 1926, with his papers on the collision processes, interpreted the wave function as a probability wave. And this interpretation, along with the uncertainty principle, is the essence of the Copenhagen interpretation. If you consider many electrons, then this probability wave becomes easy to envision: it does not reflect reality at all, but is actually just a mathematical means for predicting the relative frequency of measurement results.

Things only get tricky when you behold a single electron rather than many electrons and ask what the wave function means in that case. Does the individual electron, in its dynamic behavior, actually know something about this wave function, whose instructions in the end it somehow has to follow? The whole thing is reminiscent of radioactive decay. At the end of its lifespan, a radioactive atomic nucleus decays, and for a large group of such atomic nuclei one can easily describe how this happens, namely in the form of an exponential law. But for the individual atomic nucleus? When exactly does it fall apart? Sometimes a little earlier, sometimes a little later than the average behavior of the entire ensemble. Just like with human age: the average lifespan of a person can be specified precisely. But no one can say when the individual person will die. A decay law only applies to the decay of a whole group of radioactive nuclei, but not the individual nucleus. But which laws actually cause the decay of the individual radioactive nucleus? And the same question applies to our wave function: a wave function can make a prediction for many electrons. But what about the individual electron? Does naked, pure, natural chance rule here? And what does that actually mean: "chance"? Does the electron decide for itself, as if it had consciousness? Who governs chance? What is actually the deeper reason behind chance? Well, there is no deeper reason. A process without a deeper cause, without a deeper reason, in short a non-causal process, a process that does not follow any causal law. Today's physics accepts a non-causal description of nature: the individual electron or the individual radioactive decay is spontaneous processes that do not allow for a causal explanation. In this view, the question of what "spontaneous" means in this context is rejected as unscientific and remains unanswered.

1927, Synthesis: Dualism and Uncertainty 221

Quantum mechanics in the Copenhagen interpretation therefore establishes the invalidity of the causal law. Or better put, it establishes "conditional validity," because the Heisenberg uncertainty of measured variables becomes insignificant once we venture into the classically describable range and full causality naturally applies again. On the atomic scale, however, Heisenberg's uncertainties of quantities must be taken seriously. And chance comes into play. And the random, independent life of the electrons implies that we do not know exactly the current state of a system, i.e., position and momentum. The present is unknown. Heisenberg writes about this in his uncertainty paper:

The problem with the precise formulation of the causal law: "If we know the present exactly, we can calculate the future," is not the postscript but the premise. The premise is wrong. We principally cannot get to know the present in all its determinants. Therefore, all perception is a selection from a multitude of possibilities and a limitation of what is possible in the future.[27]

Wolfgang Pauli puts it a little differently:

Given the state of a system, generally only statistical predictions can be made about the results of future observations, while the result of individual observations is not determined by laws, but is an ultimate fact without a cause.[28]

This is said so casually: "ultimate facts without a cause." But the fact that the individual electron comes along without a law and has an effect without a cause is extremely strange. As few as a hundred electrons in a row are sufficient to reveal the pattern suggested by the wave function, but

---

[27] Heisenberg W., Über den anschaulichen Inhalt der quantentheoretischen Kinematik und Mechanik (1927, p. 197).
[28] Pauli W., Phänomen und physikalische Realität (1957, p. 43).

the single electron should only obey chance? How can the individual be random but the whole be ordered? Doesn't the single electron have to somehow, even if indirectly, obey the wave function so that a hundred electrons can actually reveal the pattern? And beyond that: "ultimate facts without a cause"—Isn't that a disclosure of the failure of physics? Didn't physics step up in the first place to find the causes of phenomena? Shouldn't the laws of physics be formulated in such a way that every observed effect can be traced back to at least one cause?

The decisive point of the Copenhagen interpretation is that chance is not the placeholder of ignorance, but reality, i.e., part of Mother Nature. The probabilistic nature of quantum theoretical predictions is not an expression of the imperfection of the theory, but rather an expression of the fact that the natural processes in the quantum world are not subject to any law of causality and have no deterministic character. The last sentence of Heisenberg's paper states almost triumphantly:

> *Rather, the true facts can be characterized much better like this: Because all experiments are subject to the laws of quantum mechanics and thus the uncertainty principle, quantum mechanics definitely determines the invalidity of the causal law.*[29]

And Pauli elsewhere:

> *The vast majority of modern theoretical physicists—especially M. Born, W. Heisenberg and N. Bohr, with whom I concur—consider these revolutionary conclusions to be irrevocable.*[30]

---

[29] Heisenberg W., Über den anschaulichen Inhalt der quantentheoretischen Kinematik und Mechanik (1927, p. 197).

[30] Pauli W., Physik und Erkenntnistheorie (1984, p. 20).

But how do the boys know this? Isn't it possible that they are simply proclaiming an imperfect theory? The appearance of chance would then simply be the result of the incompleteness of a theory that claims to be fundamental but simply is not. And this is indeed the direction that the criticism of the doubters will follow over the next few years, led by their most prominent representative: Albert Einstein.

Despite Niels Bohr's honest efforts, Como certainly does not make things any clearer. The so-called "Copenhagen interpretation" itself requires an interpretation. And it will take some time before the world of physics can really explain quantum mechanics cohesively. Bohr later sent a summary of his lecture in Como to the journal *Nature*. But there, too, confusion is so great that an editorial comment to Bohr's partly incomprehensible paper is written, conveying that quantum mechanics unequivocally cannot yet be cloaked in sufficiently clear language and that hope is, that this is not the last word in the matter.

September 1927, Somewhere in Europe

The early twentieth century is so fascinating not least because so many brilliant scientists, artists, and intellectuals were inspired by a very similar spirit. The spirit behind the new quantum mechanics was felt in the worlds of literature and philosophy just as strongly as in science. The spirit of an age can manifest itself in very different ways. And no one can say how exactly one manifestation is connected to another. You can only vaguely sense it, a certain scent in the air moving toward you.

Of all the writers who explored the astonishing work that was being done in physics at the beginning of the twentieth century, Marcel Proust is perhaps the one who most skillfully combined the poetic and the scientific. The implications of Albert Einstein's theories on the relativity of time became an important theme for Proust, who gives a stunning portrait of the subjective experience of the passing of time and memory in his epic, seven-volume novel *In Search of Lost Time*. Published over a 14-year period between 1913 and its final completion in 1927, the novel spans the life of its narrator as he grows from youth to old age, offering a spectacularly vivid portrait of French social life at the turn of the twentieth century. A masterpiece of modernist fiction, *In Search of Lost Time* breaks with nineteenth-century writing styles that centered on linear storylines and reliable narrators, focusing instead on the emergence of experience itself. Following Einstein, Proust developed a narrative style in which the experience of linear time is an illusion imposed by the mind. The most famous example of this is the "madeleine episode" right at the beginning of the first volume, in which the narrator, who is now an old man, tastes a madeleine, which immediately brings him back to the memory of eating one as a child. A memory that is completely independent of the will. We have memories that define us, but we don't know that we have them until they are awakened from their sometimes decades-old slumber by an odor or taste. Just involuntary memories! And then the present makes you realize that there is a past that suddenly shines brightly again, clearer than ever before. Time returns. Time has been found again, the time that Marcel Proust had already lost. Isn't that amazing? That there are unknown memories living inside us that somehow seem to be independent, that can be awakened completely independently of our will and only from outside? And isn't this a construction that somehow resembles the new discovery of quantum mechanics, namely that physics is determined and calculable down to a cell defined by the uncertainty of location

1927, Synthesis: Dualism and Uncertainty

and momentum, but that within this cell, chance and nothing but chance reign in complete autonomy? As we drill deeper and deeper into the mind, thoughts and memories are utterly independent and formed by pure chance. And anyone who dismisses the idea as contrived must at least admit that in both cases one is quite equally astonished that the new and unknown, following a law of its own, lies dormant and deeply embedded in the known. Proust died before the last two volumes of his work were published, but his work played an important role in a revolution in world literature that was no less exciting and important than the one taking place simultaneously in physics.

Then, let us place next to the new quantum mechanics the book "Being and Time" by Martin Heidegger, also published in 1927, a work of the century in philosophy that anticipated the transition of philosophy to existentialism by several decades. Just as is claimed for quantum mechanics, it is also said that this work could only have come about in this way due to the special circumstances in Germany in the 1920s. Heidegger's work must be placed in the context of the engagement with chaos and self-reflection that prevailed in German society after 1918. Many works appeared in German that dealt with such themes from very different directions, such as Oswald Spengler's *The Decline of the West* and Ernst Bloch's *Spirit of Utopia*, both published in 1918. The mood of many of these works is a radical break with Germany's recent past, either in the sense of an opportunity or a lament for a decline. Placing *Being and Time* in this tradition is to emphasize its radical novelty, its attempt to invent a new philosophical language to understand ourselves and its place in society. But this new language is almost incomprehensible to the layman. And quantum mechanics also develops a completely new, now mathematical language for itself: think of the operators that now represent physical quantities or the matrices that take the place of the time-dependent spatial coordinates of an orbit. In both cases, radically new forms of expression were used in order to finally be able to say what was previously unsayable.

In art too, these threads can be seen. Think of Marc Chagall, who also reached the peak of his work in 1927. Can the spirit of the time also be found in him? Is there a relationship to quantum mechanics here too? Chagall, a child of the revolution in Russia, commissar for art and culture in Vitebsk, head of an art school to which he also appointed Kazimir Malevich. Later he moved to Moscow, then to Berlin in 1922, and finally to Paris, where in the following years he worked on illustrations for Nikolai W. Gogol's "Dead Souls," La Fontaine's "Fables," and "Cirque Vollard." The years after the First World War, marked by the civil war, had not been good for Chagall, who had never really found a place in a new Russian society that he did not recognize. Chagall had been everywhere: in Palestine, Egypt, Lebanon, the Netherlands, Italy, Spain, Great Britain, and Poland, and later in the USA and Mexico. Of all the modern artists, many of whom traveled the continents in the 1920s, none left such widely scattered traces as Marc Chagall. He was internationally known for his work, but Russia, his homeland, seemed to ignore him. In order to leave a mark in his native country, he decided to leave his illustrations for Nikolai Gogol's novel at the Tretyakov Gallery in Moscow in 1927. Chagall's work recalls his early life in a small village and is rooted in Jewish identity and tradition. Unfortunately, this meant that his life in the 1930s mirrored that of many of our other protagonists. Like quantum physics, modernist art would be caught up in the wave of intolerance that spread across the continent. By staying in France, he initially avoided persecution, as did others such as Max Born and Albert Einstein. But when the Vichy regime carried out the Nazi anti-Semitic purge in 1940, he fled to New York. He was undoubtedly an international phenomenon, a man who crossed borders and was at home in the world, who was only able to find and express himself abroad. And isn't quantum mechanics also the product of a new love of internationality, of an international effort in which the habits of thought of several countries are allowed to mix?

Here, we invoke the spirit of the time, the spirit of the age, which connects works as distant from one another as those of Heidegger, Chagall, and Proust in a vague, only hinted, and more felt way with the spirit of quantum mechanics. Is it all just construction? Imagination?

What you the Spirit of the Ages call
　Is nothing but the spirit of you all,
　Wherein the Ages are reflected.[31]

Johann Wolfgang von Goethe interrupts tongue in cheek; and of course he could be right. Ultimately, it is pointless to argue about the extent to which something emerges from time and the extent to which it depends solely on creative minds. Is the zeitgeist a result of human creativity? Or is human creativity a product of the zeitgeist? Let's leave that open. In any case, quantum mechanics belongs to its time, the 1920s. That much we know.

---

[31] Goethe, Faust: The tragedy, first part, 1808. Scene: Night, Faust to Wagner.

October 1927, Leipzig

A young man approaches Heisenberg. Roughly 25 years old and thus the same age as Heisenberg. He introduces himself as Eberhard, discloses that he just saw Heisenberg play tennis and proposes a match. Werner Heisenberg had been a full professor of theoretical physics at the University of Leipzig since October 1st and a member of the Leipziger Sport-Club 1901 since October 3rd, where he was the 1201st person to be entered in the club register. Today was supposed to be his first individual lesson, but for some reason his tennis pro didn't show up. And so he plays with whoever is at hand, including this Eberhard, who he doesn't know at all, but who plays darn well. The match goes back and forth. He wins one set, narrowly, and the next one goes to Eberhard, also narrowly. And for quite some time all you can hear is the typical tennis rhythm of hitting and bouncing balls. The two men sweat and dash and reach and stretch, now chasing every ball with a calm but combative tenacity. And when the match is finally decided by a hair in favor of Heisenberg, the two exhausted combatants briefly shake hands. Eberhard, however, absolutely insists on a rematch. But Werner Heisenberg declines because he has to get to the college at the University of Leipzig. Eberhard persists and reminds him that it is just the beginning of the semester, nothing is happening at the college, the professors are just introducing themselves, everyone knows that! He could easily stay and play him again. Werner Heisenberg replies, somewhat embarrassed, that he is right and that he would very much like to play another match. The only problem is that he is not the student, but the professor. Laughing, Eberhard turns to the side: "Of course, you the professor! And I'm Henry Ford! What a lame excuse!" And walks away shaking his head.

1927, Synthesis: Dualism and Uncertainty 229

As Werner, freshly showered and walking briskly, made his way from the Sport Club 1901 to the University of Leipzig, he was truly a little irritated. He didn't seem to be able to fill the shoes of his new role as a professor. Hardly anyone took him seriously: his landlady didn't believe him and insisted on a salary verification from the University of Leipzig. And when he tried to pay a visit to his new institute at Linnéstrasse 5 for the first time on October 1st and found the main entrance locked, he encountered the caretaker, beating carpets behind the building, exclaiming: "Oh no! You are the new professor?" conveying her utter incredulity by drawing out the "you" for an insultingly long time. This, too, Werner Heisenberg endured. His father appeared dignified as a professor, so did Sommerfeld and Bohr. He, on the other hand, looked like a farm boy who had thrown on his Sunday best for a trip to town.

He did not care what his landlady or the building caretaker thought of him, whether they acknowledged his professorial dignity. Irrelevant. What concerned him far more was the fact that only two students had decided to take his course. Would students take him seriously? Would he be able to build a group here in Leipzig that would be recognized internationally? He hadn't had to worry about winning over students until now. He had always been the one being courted, never the one doing the courting. When he began his studies in Munich exactly 7 years ago, Arnold Sommerfeld supervised him intensively almost from day one. Then there was Max Born, who practically adored him. And then the years with Niels Bohr! My God, how much his elders had cared about him, how much they encouraged him. But that was over now. Now he was the elder!

He had reached Linnéstrasse 5 and bounced up the stairs to the institute entrance. He had an idea: he would organize a student exchange with Wolfgang Pauli's group. Pauli would soon be starting in Zurich and his students would certainly find an exchange there attractive. Maybe new students can be recruited this way. Things will work out, Heisenberg thinks, things will work out!

October 1927, Cambridge

St John's College is the richest college in Cambridge, but in 1927 it could not yet boast of a Nobel Prize winner among its fellows. But now Paul Dirac is a fellow at St. John's and in pursuit of one. Paul Dirac, son of an immigrant French teacher from Bristol and now a fellow at the venerable St. John's College. From rags to riches: is that even fathomable?

In previous years, the college porter had never greeted Paul or even looked up when he handed him the key to his room. But today, as he showed him to his little apartment at the college, he was a changed man, barking "Yes, sir!" in military style, conveying his readiness to serve. And once the ridiculous porter had left the room and Paul was watching the newly enrolled students down at the Tudor Gate through the window, his personal miracle suddenly became clear to him. Exactly four years ago, on October 1, 1923, he himself had stood down there at the Tudor Gate. And just two years ago, before he got his hands on the proofs of Heisenberg's Heligoland article, he had been an inconspicuous and completely insignificant student. And today, 2 years later, he was a Fellow at St. John's College! Was one of 29 physicists invited to the fifth Solvay Conference in Brussels. The big summit meeting, to which only the crème de la crème were invited. And personalities such as Max Born, Erwin Schrödinger, Niels Bohr, and Albert Einstein spoke of his scientific achievements with reverent respect.

Paul Dirac turned away from the window and sat down at his still-empty desk. He would present his quantum field theory at the Solvay Conference, unsure but also not worried how that might turn out. And upon returning from Brussels, he would do what he had wanted to do for a long time. Right here, at this desk, he would get serious. No, he had not yet reached his professional climax, there was more to come. Paul Dirac was absolutely sure. There was more to come. Here at this desk, he would melt relativity and quantum mechanics into a common theory. He would not leave this feat of glory to Wolfgang Pauli. Nor to Pascual Jordan or Oskar Klein. This time he would not be late, this time he would hit the big time, there was no doubt. And once accomplished, he would write a book, also here at this desk. A textbook on quantum mechanics in which all the great and unique things that he had seen and learned to understand in recent years would be laid out in simple, clear, and straightforward language. Yes, thought Dirac, that's what will happen. I created quantum mechanics, and quantum mechanics created me.

October 1927, Brussels

Dear Sir!

On behalf of the Scientific Committee of the "Institut International de Physique Solvay", I have the pleasure of inviting you to the conference that will take place in Brussels from October 24th to approximately 29th. It will be devoted to the new quantum mechanics and related questions, and we will have reports from Messrs. Born and Heisenberg, Bragg, de Broglie, Compton and Schrödinger to introduce the discussion. It would be a great pleasure for us if you could take part in this conference. I would be very happy to receive a positive answer.

With the utmost respect

Yours sincerely, H.A. Lorentz[32]

Anyone who found this letter in their mailbox in mid-September had a probability of more than 50% of receiving a Nobel Prize: of the 29 invited conference participants, 17 received a Nobel Prize in their lifetime. This fifth Solvay Conference in Brussels is probably the most famous conference in the history of physics. At the same time, it marks a turning point: the end of one scientific epoch and the start of the next. The new quantum theory could finally be considered understood. This was the end. And it was the beginning of what Heisenberg called the "golden age of atomic physics." Heisenberg in "The Part and the Whole":

The gates to the newly opened field of quantum mechanics of the atomic shell were wide open; and those who wanted to research and work here, who wanted to pick the fruits of the garden, were faced with countless problems that were previously unsolvable and could be dealt with and resolved using the new method.[33]

At the same time, the fifth Solvay Conference began a debate that lasted for many years based on the two giants Niels Bohr and Albert Einstein: the so-called Bohr-Einstein debate. The intensive discussions between the two physicists characterized the days in Brussels and made the viewer realize that although the quantum theory was finished, there was something unsatisfactory about it. One camp interpreted this as a feature of nature and the other camp interpreted it as an indication that quantum theory is incomplete and therefore only a provisional theory.

---

[32] Lorentz to Pauli on September 5, 1927, see Klein & Toomer (1979, p. 408).
[33] Heisenberg W., Der Teil und das Ganze (2022, p. 114).

1927, Synthesis: Dualism and Uncertainty

The dispute led by Messrs. Bohr and Einstein once again reveals the essence of quantum mechanics. The wonder of quantum mechanics is best illustrated by the double slit experiment, although this experiment was only discussed after the fifth Solvay Conference. A wave striking a double slit results in a striped pattern on a screen behind the double slit. This interference pattern can be viewed as a sign of the wave. So if you want to know whether the electron is a particle or a wave or both, the best thing to do is send electrons to such a double slit. If the electron also has a wave character, this characteristic interference pattern should somehow be visible behind it. Let's go! Let a beam of electrons hit a double slit. The electrons approach the double slit one at a time. Behind the double slit is a screen where the electrons strike and are detected. The first impacts lead to points that appear to be completely randomly distributed across the screen. After a few hundred impacts, however, the distribution of the many electron impacts on the screen evolves into the typical interference pattern, i.e., they are distributed as one would expect for a wave influenced by a double slit. The wave pattern corresponds entirely to the predictions of Schrödinger's wave function, and all is well with the world. Quantum mechanics can predict the distribution of measurement results well.

Things only get really strange when you look at the individual electron and its path through the apparatus. In order for many hundreds of electrons to later show the characteristic stripe pattern in their distribution, the individual electron, despite its seemingly random behavior, must somehow take into account the underlying wave. And this is where things get weird. Please try to imagine: the path of a single electron through the double slit. Although the electron necessarily only passes through one slit, it somehow follows the directions of the accompanying wave, which passes through both slits at the same time and then forms the characteristic interference pattern behind it. How can one actually imagine this dualism of the electron and its wave? How is it possible for an electron to go through one slit and still react to the other slit? It is almost as if the electron had an "awareness" of the other slit. And, yes, this word from psychology can also be found in the physical literature on the double slit. A certain Mr. Cochran wrote in 1971:

> *Each electron passes through only one slit but is aware of the existence and location of the other slit if it is open, and chooses different diffraction angles when this second slit is open—angles that allow it to produce part of the characteristic diffraction pattern. Rather than being something that has both particle and wave properties, the electron in this concept is a particle that has a degree of consciousness.*[34]

The Copenhagen School makes it easy here. In Heligoland, Heisenberg had said goodbye to the idea of an electron orbit and instead, with almost brutal consistency, had adopted the idea that only things that can be measured should be described theoretically. In this respect, quantum mechanics no longer provides an electron path, but only a prediction for the electron distribution on the screen behind the double slit. We can only speculate how exactly the electron gets through one slit and at the same time is able to react to the other slit; in any case, that lies outside the domain covered by quantum theory. The physicists from the Copenhagen School simply do not consider themselves responsible for this issue. Working out or even allowing a more precise idea of the dualism of particles and waves is not part of the Copenhagen program.

---

[34] Cochran (1971, p. 245).

1927, Synthesis: Dualism and Uncertainty 235

But there are also "the others": Albert Einstein, Max Planck, Paul Ehrenfest, Louis de Broglie, Erwin Schrödinger, and many more. They want to be able to imagine things beyond the measurable. They want to understand the dualism of particles and waves in more detail. Above all, they do not accept that chance should be a fundamental principle of Mother Nature and that there should be non-causal processes in the atomic world. Imagine the coincidence visually: how the electron decides for itself in which direction it moves, which slit it takes, where it jumps and leaps away. And this is where their arguments begin. Einstein wrote about this in 1924:

> *In contrast to Bohr, I do not want to be driven to abandon strict causality until we have defended ourselves against it in a completely different way than before. The idea that an electron exposed to a beam freely chooses the moment and direction in which it wants to jump away is unbearable to me.*[35]

And elsewhere Carsten Held attributes the following considerations to Einstein:

> *Quantum jumps, such as quantum light absorption and emission and the "jump away" of the electron, are phenomena that actually require explanation in atomic physics. [....] The task of physics is to explain them. And the shortcoming of quantum mechanics is that it does not provide this, but rather that it dispenses with any explanation in order to then make a virtue out of this renunciation.*[36]

So this is how Albert Einstein sees it, and Albert Einstein is not just anyone. He was the one who introduced the

---

[35] Born M., Einstein-Born. Briefwechsel 1916–1955. third Edition (2005, pp. 140–1).
[36] Held (1999, pp. 73–4).

concept of photons and thus assigned the particle character to the light wave and who spent his entire life thinking about the questions of the dualism of particles and waves. Einstein, as spokesman for the opposition, essentially says that quantum mechanics simply does not explain the phenomenon that actually needs to be explained. The fact that quantum theory has a statistical character, i.e., operates with probabilities, is not an expression of a fundamental property of nature, but rather just a sign of an incomplete theory. Einstein's most famous quote can be found in a letter to Max Born:

> *I find your statistical interpretation of the wave function very respectable. But an inner feeling tells me that this isn't the real deal after all. The theory provides a lot, but it hardly brings us any closer to the secret of the old man. In any case, I'm convinced that he doesn't roll dice.*[37]

The good Lord doesn't throw dice. This is Einstein's famous saying. He means that chance as an elementary phenomenon of nature is simply inconceivable. How on earth is such a coincidence accomplished? How is that supposed to work? Doesn't this almost reflexively lead to the rather questionable idea that the electron has consciousness and decides paths for itself? Einstein and his followers are much more likely to assume that this chance is entirely reducible and that it can somehow be traced back to underlying causes.

So now they meet in Brussels for the Solvay Conference. And Albert Einstein has intently prepared for this discussion. He hasn't been invited to lecture, but he doesn't care. Niels Bohr once again gives his philosophical

---

[37] Letter from Einstein to Max Born dated December 4, 1926 (Born M., Einstein-Born. Briefwechsel 1916–1955. third Edition, 2005, p. 154).

1927, Synthesis: Dualism and Uncertainty 237

complementarity lecture from Como. And when he is finished, Albert Einstein immediately starts asking questions. The sucker punch was that he didn't address Niels Bohr and his notions at all, probably because he had no idea what to do with them and didn't see the point of the presentation. So he ignored Bohr's considerations, a fact that surely hurt his feelings. But this is what got the conference really going, because then a genuine discussion kicked off that would continue for years to come. Heisenberg reports:

We all lived in the same hotel, and the most heated discussions took place not in the conference room but during meals in the hotel. Bohr and Einstein did most of the heavy lifting in this battle over the new interpretation of quantum theory. Einstein was unwilling to accept the fundamentally statistical nature of the new quantum theory. Of course, he had no objections to using probability statements whenever all the details of the system in question were not known. The earlier statistical mechanics and thermodynamics were based on such statements. However, Einstein did not want to concede that it should be fundamentally impossible to know all the parameters necessary for a complete determination of the processes. "The good Lord doesn't play dice," was a phrase you often heard him use in these discussions. Therefore, Einstein could not come to terms with the uncertainty relations and he tried to think of experiments in which these relations no longer hold.[38]

The fact that the entire Solvay Conference revolved primarily around Heisenberg's uncertainty principle contributed significantly to its phenomenal fame. Among all the abstractions that this difficult theory had produced, the uncertainty principle stood out as a truly tangible and imaginable result. One could engage in it. In this, and with this, uncertainty principle, Heisenberg distilled the essence

---

[38] Heisenberg W., Der Teil und das Ganze (2022, p. 99).

of quantum mechanics. This was the demarcation from classical physics that highlighted the secret workings of chance. And everyone present realized what Heisenberg had already written in his uncertainty paper:

> *If there were experiments that simultaneously enabled a "sharper" determination of p and q than corresponds to the uncertainty principle, quantum mechanics would be impossible.*[39]

Einstein knew that too. And so Einstein wanted to achieve exactly that: to find just one experiment that disproves the uncertainty principle, then the entire quantum mechanics would be refuted. If you crack that, you crack everything. Let's listen further to Heisenberg's conference report:

> *The arguments usually began early in the morning when Einstein explained to us a new thought experiment for breakfast that, in his opinion, refuted the uncertainty relations. Of course, we began the analysis immediately, and on the way to the conference room an initial clarification of the question and the claim was achieved. There were many discussions about it over the course of the day, and usually in the evening it was time for Niels Bohr to prove to Einstein over a shared meal that even the experiment he proposed could not lead to a circumvention of the uncertainty relations.*

It is beautifully apparent what an extraordinary role Einstein played during this gathering. Everyone looked up to him. Every physicist present had grown up with Einstein's Theory of Relativity. He was the final authority. And now this final authority simply did not want to bless the new

---

[39] Heisenberg W., Über den anschaulichen Inhalt der quantentheoretischen Kinematik und Mechanik (1927, pp. 179–180).

theory. So the warriors of quantum mechanics gathered behind Nils Bohr to prepare to storm the bastion.

Einstein was a little perturbed, but by the next morning at breakfast he had a new thought experiment ready, more complicated than the previous one, which was supposed to demonstrate the invalidity of uncertainty relations once and for all. Of course, this attempt ultimately didn't fare any better than the first, and after this game had continued a few times, Einstein's friend Paul Ehrenfest, a physicist from Leyden in Holland, said: "Einstein, I'm ashamed of you; Because you are arguing against the new quantum theory in just the same way as your opponents argue against the theory of relativity." But even this friendly admonition could not convince Einstein.[40]

But why should Einstein stop when he simply wasn't convinced yet? This quarrel, and that's what made it so exciting, actually revolved more and more around the question of whether, behind the world described by quantum mechanics, there is a deeper world whose description, which has yet to be found, no longer allows for chance and in fact restore the validity of full causality. Heisenberg put it this way in his uncertainty paper:

Since the statistical character of quantum theory is so closely linked to the inaccuracy of all perception, one could be tempted to assume that behind the perceived statistical world there is still a "real" world in which the causal law applies. But such speculation, and we must emphasize this explicitly, seem fruitless and pointless. Physics is meant to formally describe only the connection between perceptions.[41]

---

[40] Heisenberg W., Der Teil und das Ganze (2022, pp. 99–100).
[41] Heisenberg W., Über den anschaulichen Inhalt der quantentheoretischen Kinematik und Mechanik (1927, p. 197).

This is exactly the basic attitude of the so-called "positivists", whose principle is: Only what can be perceived is real. And physics alone is supposed to explain the connections between perceptible phenomena. Questions, on the other hand, that cannot be decided through observation and measurement do not belong to the business of physics. They are pointless metaphysical issues. Of course, it's no wonder that Heisenberg thought this way. The radical focus on the measurable was the key to his success: after all, his work on Heligoland began with the resolute abandonment of the concept of an electron orbit, precisely because it cannot be measured at all. Pauli sums up the positivist approach:

Material or general physical objects whose nature is supposed to be independent of the way in which they are observed are metaphysical extrapolations. We have seen that modern physics, forced by facts, had to abandon this abstraction as too narrow.[42]

At the fifth Solvay Conference, the realist school separated from the positivist school. The two schools can be distinguished by the answer to the question: Do the fundamental objects of atomic physics—electrons, photons, and the atoms themselves—exist independently of humans, their measurements and observations? The realists behind their leader Einstein answer this question with "Yes!" and the positivists answer: it doesn't matter at all! Heisenberg:

However, all opponents of quantum theory agree on one point. In their view, it would be desirable to return to the concept of reality of classical physics, or more generally, to the ontology of materialism, i.e. to the idea of an objective, real world whose smallest parts exist objectively in the same way as stones and trees, regardless of whether we watch them or not.[43]

---

[42] Pauli W., Physik und Erkenntnistheorie (1984, p. 16).
[43] Heisenberg W., Physik und Philosophie (1959).

1927, Synthesis: Dualism and Uncertainty 241

Einstein's entire research life presumed such an objective world in which physical processes occur according to fixed laws, independent of measurement and observation. According to Einstein, the mathematical symbols of theoretical physics reflect this objective world and allow predictions of future behavior. And now it is claimed that there is no such objective world at the atomic level and that the mathematical symbols only represent the possible, not the factual. Preposterous!

According to Heisenberg, the world comes into existence only through measurement. According to Einstein, however, first comes the world and then measurement. The world precedes all empirical knowledge. Einstein in 1931:

The belief in an external world independent of a perceiving entity underlies all natural science.[44]

This is the realist speaking. The positivist Pascual Jordan, who was not invited to Brussels despite his substantial contributions to quantum mechanics, formulates the opposite position:

The common misunderstanding is that the positivist approach denies the existence of a real external world. However, the negation of a meaningless statement results in a meaningless statement; The assertion of the non-existence of a real external world is therefore no more meaningful than the assertion of its existence. Both are neither right nor wrong, but meaningless.[45]

The Solvay Conference closed with an open ending. Right until the end, Einstein had not found a really convincing example that could disprove the uncertainty

---

[44] Einstein A., Mein Weltbild (1993, p. 159).

[45] Jordan P., Anschauliche Quantentheorie (1936, p. 303).

principle. The young quantum mechanics had passed its first stress test. And Einstein found himself on the defensive.

By quoting Paul Ehrenfest and his shame, Heisenberg makes it seem as if Einstein was just a stubborn old man who begrudged the young people their success. But that's rather unfair. Firstly, Einstein's obvious question is whether it might not be possible to complete the theory in such a way that it would then become deterministic again.

And secondly, Einstein's objections cannot have been entirely nonsensical, given that they kept physics in suspense for many decades to come. Initially, in 1932, John von Neumann succeeded in finding a theorem which seemed to prove that a causal extension to quantum mechanics was impossible. It later became clear that the Von Neumann theorem was not really applicable. Then in 1935 came the Einstein-Podolsky-Rosen thought experiment and once again there was great excitement because Einstein and his colleagues had discovered that the quantum entanglement permitted by quantum mechanics would lead to a violation of the classical principle of local realism. The term "quantum entanglement" originated with Schrödinger, also in 1935. When two particles are in entangled quantum states, the measurement result from one particle automatically fixes the result of an equivalent measurement of the other particle, even if the other particle is far away. And again, one could get the impression that quantum mechanics violated fundamental principles. Maybe there was a complete theory beneath quantum mechanics after all. This is what de Broglie wrote in 1953.

> *The question is whether the now accepted interpretation of quantum mechanics is incomplete and hides a completely*

1927, Synthesis: Dualism and Uncertainty

deterministic reality that can be described in space and time by hidden variables.[46]

And in 1975 Paul Dirac said:

In my opinion, it is likely that at some point in the future we will have improved quantum mechanics, which will mean a return to determinism and thus vindicate Einstein's views.[47]

But here the great Dirac was completely mistaken. In 1964, the Irishman John Stewart Bell came along with his "inequality" concept and then in 1974 a certain John Clauser from Berkeley, who tested Bell's inequality and thus finally—because he did so experimentally—settled the dispute that began at the fifth Solvay Conference in Brussels: Einstein had believed in local realism in vain. Heisenberg and the Copenhagen School had been right: there is *indeed* a fundamental uncertainty in the measurements, there is *indeed* chance as a natural principle, and one must *indeed* bid causality farewell. Quantum mechanics is not a provisional, but a *fundamental* theory. For this, John Clauser along with others received the Nobel Prize in 2022.

And even if it was anything but obvious at the time, the fifth Solvay Conference actually marked the end of the origins of quantum mechanics. What had begun in Heligoland came to an end in Brussels.

---

[46] Selleri (1990, p. 49).
[47] Ibid., p. 193.

Epilogue: The Eight Protagonists in the Garden Hall

It has turned dark in the large garden hall, the sky is already largely pitch black, only the horizon still shows a faint glow, against which the silhouettes of the trees stretch toward the sky. Of the eight invited guests, one has already left, Niels Bohr. When talk turned to his argument with Heisenberg, he jumped up angrily, asked for his hat and coat, and left the gathered party prematurely. He deems it simply unacceptable to fault him in this fashion for his grappling and his searching. Everyone makes mistakes in cases like this. Yes, he knows that it was only in the 1930s that he managed to formulate the principle of complementarity in a truly coherent way. But he deserved more respect and so his early departure from the group ought to be seen as a protest. Everyone glances at Werner Heisenberg, but he glances at the ground. When this tale was truly over, Wolfgang Pauli is the first to say goodbye. He, too, is cross. He doesn't understand why people always criticize him for his objective superiority. Surely everyone in this room knows that it was essentially his ideas that were the real fuel behind this story.

Epilogue: The Eight Protagonists in the Garden Hall

Schrödinger raises his eyebrows and picks up his coat as well. With Erwin Schrödinger and Louis de Broglie already at the door, you can still hear de Broglie's voice out of the dark: he points out that he gave a really good talk at the Solvay Conference and offered an interpretation of quantum mechanics that would still be relevant many years later. Why was that not properly showcased in this play? And does he, Schrödinger, really think he was a theoretical bungler. Schrödinger turns to him with a laugh: "I don't think you are the only theoretical bungler!" Max Born, on the other hand, leaves the evening in a brilliant mood. He had had a good time. And when Werner Heisenberg remarked as he got ready to leave that a lot of things had been made up, for example, that he had never been a member of that tennis club in Leipzig, Born patted him on the back with a laugh and said that he, Heisenberg, should be the first to appreciate a little uncertainty in the story. Only unimportant things were fabricated, and only to make it a better story. Whether he, Heisenberg, had actually ever even been in this garden room. No, of course not, Heisenberg admits. There you go, says Born, gets into the car, waves and drives off. Jordan and Dirac are the only ones left in the garden hall. Pascual Jordan requested a bottle of cognac because he wanted to sit by himself for a while. He would then turn off the lights and show himself out. But Paul Dirac beholds the unhappy Jordan for a long time. As he says goodbye, he mumbles sheepishly toward him that everyone has their destiny and that in the end no-one can escape it. Then he slowly walks through the park to the gate. You can faintly hear the crunching sound of footsteps on the pebble path. Then all is still.

In 1932, Paul Dirac became a mathematics professor at the University of Cambridge. He taught there until 1969, before he went to Florida for a few years. Louis de Broglie

Epilogue: The Eight Protagonists in the Garden Hall

was appointed professor of theoretical physics in 1928, first at the Henri Poincare Institute, and then at the Sorbonne in 1932, where he taught until 1962. Wolfgang Pauli was appointed to ETH Zurich in 1928, but in the 1930s he worked for increasingly long periods in the USA, primarily at the Institute for Advanced Study, a private research institute founded in Princeton in 1930 and where Albert Einstein had also worked. In the spring of 1933, Max Born was forced to take a leave of absence by the Nazis because of his Jewish origins. He emigrated to England, first to Cambridge, then went to the University of Edinburgh in 1936 before returning to Germany in 1953. And it was already clear to Erwin Schrödinger in the first half of 1933 that he did not want to stay in Germany. He simply never returned to Berlin from his Tyrolean summer vacation and then spent three unhappy years in Oxford, where he was a fellow at Magdalen College. In the fall of 1936, he took over a professorship at the University of Graz and was dismissed from this professorship without notice by the Nazis on March 31, 1938, because of political unreliability. He then went to Dublin for 17 years, only returning to the University of Vienna in 1956. Pascual Jordan received a professorship at the University of Rostock in 1929, and in 1944 he was appointed to the University of Berlin as Max von Laue's successor. He later sat as a member of the CDU in the German Bundestag. Niels Bohr stayed in Copenhagen and was involved in the resistance during the German occupation of Denmark. In 1943, he fled to Sweden and then traveled to the USA under the code name Nicholas Baker, but returned to Denmark immediately after the war. Finally, Werner Heisenberg: Together with Friedrich Hund, he turned his chair at the University of Leipzig into a center for theoretical physics. In 1942, he was put in charge of the Kaiser Wilhelm Institute for Physics in Berlin Dahlem.

Epilogue: The Eight Protagonists in the Garden Hall

Heisenberg was involved in the Army Weapons Office's uranium project. After his internment at Farm Hall, he became one of the leading scientists in the young Federal Republic of Germany.

Timeline

1924

22 January: Ramsay McDonald becomes the first Labor Prime Minister in the United Kingdom, the first from the working class.

February: In India, Gandhi is released from prison after serving 2 years for sedition against British rule.

1 April: Adolf Hitler is sentenced to prison for his role in the 1923 putsch. He only served nine months.

6 April: Benito Mussolini's fascist party comes to power and wins a two-thirds majority in the election.

May: Together with Kramers and Slater, Niels Bohr publishes the article *The quantum theory of radiation*. Discussions begin about the Bohr-Kramers-Slater theory (BKS theory), one of the important bridges from classical physics to quantum mechanics.

28 June: Heisenberg successfully defends his habilitation thesis in Göttingen; receives the venia legendi in physics from the University of Göttingen in October.

16 August: The Dawes Plan is signed in Paris, which ends the French occupation of the Ruhr area and reorganizes Germany's reparation payments.

17 September: Werner Heisenberg moves to Copenhagen to work with Niels Bohr at his famous institute.

23 November: Edwin Hubble publishes his findings in the *New York Times* that the universe is actually larger than previously thought.

November: Louis de Broglie defends his doctoral thesis "*Research on quantum theory,*" Pascual Jordan publishes his first scientific article in the *Zeitschrift für Physik*.

1925

1 January: Pascual Jordan takes up his position as assistant to Max Born in Göttingen.

3 January: Benito Mussolini is declared Dictator of Italy.

16 January: The *Zeitschrift für Physik* receives Wolfgang Pauli's article on the exclusion principle.

28 February: Reich President Friedrich Ebert dies in office.

March: Paul Dirac spends a short time with his parents in Bristol after his brother Felix commits suicide and then returns to Cambridge.

10 April: F. Scott Fitzgeralds novel *The Great Gatsby* is published.

5 May: Teacher John Scopes is arrested in Tennessee for teaching Darwin's theory of evolution. A turbulent trial begins that is followed worldwide.

12 May: Niels Bohr visits Cambridge and meets Paul Dirac for the first time.

7 June: Heisenberg suffers from severe hay fever and retreats to the island of Heligoland, where he works on a paper on new quantum mechanics.

29 July: Submission of Heisenberg's seminal article "*On quantum theoretical reinterpretation of kinematic and mechanical relationships*" at the *Zeitschrift für Physik*. The beginning of an extraordinary series of papers by Heisenberg, Born and Jordan developing Göttingen matrix mechanics.

July/August: Heisenberg travels to Leiden and Cambridge and meets, among others, Ralph Fowler.

August: Paul Dirac receives the proofs of Heisenberg's Heligoland article, which prompts him to delve deeper into the new quantum theory.

27 September: Receipt of the article "*On quantum mechanics*" by Max Born and Pasucal Jordan at the *Zeitschrift für Physik*.

29 October: Max and Hedi Born embark on the ship *Westphalia* and begin a lecture tour through the USA that introduces the new quantum mechanics to America.

7 November: Receipt of paper "*The Fundamental Equations of Quantum Mechanics*" by Paul Dirac at the magazine *Proceedings of the Royal Society*.

16 November: Entrance of the "Three Men's Work" under the title "*On quantum mechanics II*" at the *Zeitschrift für Physik*.

20 November: In issue 47 of the magazine "*Natural Sciences*" appears the preliminary message "*Replacing the hypothesis of non-mechanical constraint with a requirement regarding the internal behavior of each individual electron*" by G. Uhlenbeck and S. Goudsmit, which introduces the concept of spin.

1 December: The Locarno Accords are signed, normalizing international relations with Germany.

1926

26 January: Scottish inventor John Logie Baird demonstrates the first working television.

27 January: Receipt of the article "*Quantization as an eigenvalue problem (first communication)*" by Erwin Schrödinger at the magazine *Annals of Physics*. This begins a series of four articles ("four communications") in this journal by which Schrödinger develops his wave mechanics as an alternative formulation of quantum mechanics.

8 February: Greta Garbo makes her Hollywood debut in the silent film *Torrent*. She later becomes one of the most famous actresses of Hollywood's Golden Age.

21 April: Queen Elizabeth II is born in London. After the death of her father George VI, she ascended the throne on February 6, 1952, and reigns until her death in 2022, the longest reign of any British monarch.

1 May: Beginning of Werner Heisenberg's employment as a university lecturer and assistant to Niels Bohr in Copenhagen.

3 May: A general strike begins in the United Kingdom with 1.7 million striking workers, the only general strike in the history of the UK. Paul Dirac submits his doctoral thesis entitled "*Quantum Mechanics.*"

June: Pascual Jordan travels to Bad Pyrmont to receive treatment for his stuttering, paid for by Niels Bohr. Paul Dirac receives his doctorate from Cambridge.

21 July: Max Born sends his essay "*Quantum mechanics of collision processes*" to the *Zeitschrift für Physik* after he had already sent a preliminary notice there in June. Pascual Jordan travels to Vienna to be treated by the famous psychotherapist Alfred Adler.

16 July: Erwin Schrödinger explains his theory in a lecture at the German Physical Society and on July 17th at the Physics Colloquium at the University of Berlin.

23 July: Schrödinger speaks in the Audimax of the Ludwig-Maximilian University of Munich and meets Werner Heisenberg for the first time.

26 August: Receipt of the article "*On the Theory of Quantum Mechanics*" by Paul Dirac at the Royal Society, in which Fermi-Dirac statistics are elaborated as an integral part of quantum mechanics.

1 September: Paul Dirac begins his postdoctoral period, initially with a stay of several months at Bohr's institute in Copenhagen.

1 October: Erwin Schrödinger visits Niels Bohr in Copenhagen for a few days.

1 November: Wolfgang Pauli becomes a professor at the University of Hamburg. Pascual Jordan completes his habilitation thesis at the age of just 24.

December: Paul Dirac and Pascual Jordan independently send their articles to the Royal Society and the *Zeitschrift für Physik*, which together make up the Dirac-Jordan transformation theory and represent the final coherent mathematical basis for quantum mechanics.

1927

10 January: Fritz Lang's classic film *Metropolis* debuts, an iconic work of German expressionist cinema.

27 January: Dirac travels to Göttingen to work with Max Born's group. He shares his apartment in Göttingen with Robert Oppenheimer.

23 February: Werner Heisenberg writes a letter to Wolfgang Pauli in which he presents his thoughts on the uncertainty principle for the first time.

23 March: Introduction to Heisenberg's work on the uncertainty principle in *Zeitschrift für Physik*.

April: Belgian astrophysicist and Catholic priest Georges Lemaitre first proposes the idea of a "big bang" as the origin of our universe, as well as the idea that the universe expands and eventually contracts.

20–21 May: The American pilot Charles Lindbergh (1902–1974) flies over the North Atlantic in a west-east direction non-stop and alone in 33.5 hours in his single-engine "Spirit of St. Louis."

23 July: The architectural exhibition "The Apartment" is opened by the German Werkbund in Stuttgart. The *Weissenhof* housing project with 60 residential units, designed for the exhibition by multiple leading European architects, is completed under the direction of Ludwig Mies van der Rohe.

11 September: Volta Conference in Como. Niels Bohr presents his principle of complementarity and formulates the "Copenhagen interpretation" of quantum mechanics for the first time.

1 October: Schrödinger takes over the chair of theoretical physics at the University of Berlin from Max Planck.

1 October: Heisenberg is appointed and takes over the chair for theoretical physics at the University of Leipzig. Paul Dirac becomes a Fellow of St. John's College, Cambridge.

24–29 October: The fifth Solvay Conference takes place in Brussels. It brings together almost all of the protagonists in the story on the origins of quantum mechanics, which formally comes to an end here. At the same time, a decades-long discussion about the implications of this theory begins.

Bibliography

Bohr, M. (1962, November 7). Interview of Niels Bohr Session III. (T. S. Kuhn, L. Rosenfeld, A. Petersen, & E. Rudinger, Interviewers) Niels Bohr Library & Archives, American Institute of Physics. Retrieved from www.aip.org/history-programs/niels-bohr-library/oral-histories/4517-3

Born, M. (1926a). Quantenmechanik der Stoßvorgänge. *Zeitschrift für Physik, 38*, 803–827.

Born, M. (1926b). Zur Quantenmechanik der Stoßvorgänge. *Zeitschrift für Physik, 37*, 863–867.

Born, M. (1978). *Mein Leben.* München: Nymphenburger Verlagsbuchhandlung.

Born, M. (2005). *Albert Einstein, Max Born. Briefwechsel 1916–1955. 3rd Edition.* München: Lange Müller.

Born, M. (2015). *My Life: Recollections of a Nobel Laureate.* London: Taylor & Francis.

Born, M., & Jordan, P. (1925a). Zur Quantenmechanik. *Zeitschrift für Physik, 34*, pp. 858–888.

Born, M., & Jordan, P. (1925b). Zur Quantentheorie aperiodischer Vorgänge. *Zeitschrift für Physik, 33, 479.*

Cassidy, D. C. (1995). *Werner Heisenberg, Leben und Werk.* Heidelberg, Berlin: Spektrum Akademischer Verlag.

Cochran, A. A. (1971). Relationship between quantum physics and biology. *Found. Phys, 1*, pp. 235–250.

Dahn, R. (2019). *The Forgotten Founder of Quantum Mechanics: The Science and Politics of Physicist Pascual Jordan, 1902–1980.* Chicago: University of Chicago.

Dirac, P. (1962, April 1). Interview of P. A. M. Dirac, Session I. (T. S. Kuhn, & E. Wigner, Interviewers) American Institute of Physics, College Park, MD USA: Niels Bohr Library & Archives. Retrieved from www.aip.org/history-programs/niels-bohr-library/oral-histories/4575-1

Dirac, P. (10. May 1963). Interview of P. A. M. Dirac, Session IV. (T. S. Kuhn, Interviewer) American Institute of Physics, College Park, MD USA: Niels Bohr Library & Archives. www.aip.org/history-programs/niels-bohr-library/oral-histories/4575-4

Dirac, P. A. (1925). The fundamental equations of quantum mechanics. *Proc. R. Soc, A109*, 642–653.

Einstein, A. (1922). Besprechung des Enzyklopädieartikels von Wolfgang Pauli. *Die Naturwissenschaften, 10*, pp. 184–5.

Einstein, A. (1979). *Aus meinen späten Jahren.* Stuttgart: DVA.

Einstein, A. (1993). *Mein Weltbild. (C. Seelig, Hrsg.).* Frankfurt am Main, Berlin: Ullstein (25. Auflage).

Elsasser, W. (1925). Bemerkungen zur Quantenmechanik freier Elektronen. *Naturwissenschaften, 13(33)*, p. 711.

Enz, C., & von Meyenn, K. (1988). *Wolfgang Pauli, Das Gewissen der Physik.* Braunschweig: Vieweg.

Farmelo, G. (2009). *The Strangest Man.* London: Faber & Faber.

Farmelo, G. (2016). *Der seltsamste Mensch.* Berlin, Heidelberg: Springer.

Fischer, E. P. (2022). *Die Stunde der Physiker.* München: C.H. Beck.

Greenspan, N. T. (2006). *Max Born, Baumeister der Quantenwelt.* Heidelberg: Spektrum Akademischer Verlag.

Heisenberg, W. (1971). *Schritte über Grenzen, Gesammelte Reden und Aufsätze.* München: Piper Verlag.

Heisenberg, W. (1925). Über quantentheoretische Umdeutung kinematischer und mechanischer Beziehungen. *Zeitschrift für Physik, 33*, pp. 879–893.

Heisenberg, W. (1927). Über den anschaulichen Inhalt der quantentheoretischen Kinematik und Mechanik. *Zeitschrift für Physik, 43*, pp. 172–198.

Heisenberg, W. (1959). *Physik und Philosophie.* Frankfurt: Ullstein.

Heisenberg, W. (1963a, July 5). Interview of Werner Heisenberg. (T. S. Kuhn, & J. L. Heilbron, Interviewers) College Park, MD, USA: Niels Bohr Library & Archives, American Institute of Physics. Retrieved from www.aip.org/history-programs/niels-bohr-library/oral-histories/4661

Heisenberg, W. (1963b, July 5). Interview of Werner Heisenberg. College Park, MD USA: Niels Bohr Library & Archives, American Institute of Physics. www.aip.org/history-programs/niels-bohr-library/oral-histories/4661. (T. S. Kuhn, & J. L. Heilbron, Interviewers)

Heisenberg, W. (2022). *Der Teil und das Ganze.* München/Berlin: Piper Verlag.

Held, C. (1999). *Die Bohr-Einstein-Debatte: Quantenmechanische und physikalische Wirklichkeit.* Paderborn: Mentis Verlag.

Hermann, A. (1976). *Heisenberg.* Reinbek bei Hamburg: Rowohlt Taschenbuch Verlag.

Hirsch, M. (2003). *Werner Heisenberg - Liebe Eltern. Brief aus kritischer Zeit. 1918–1945.* München: Langen-Müller.

Hoffmann, D. (1984). *Erwin Schrödinger (Bd. 66).* Leipzig: BSB B.G. Teubner Verlagsgesellschaft.

Hürter, T. (2021). *Das Zeitalter der Unschärfe.* Stuttgart: Klett-Cotta.

Jordan, E. P. (1963, June 17, 18, 19 und 20). Interview. (T. S. Kuhn, Interviewer) American Institute of Physics, College Park, MD USA: Niels Bohr Library & Archives. Retrieved from http://repository.aip.org/islandora/object/nbla:269316

Jordan, P. (1924). Zur Theorie der Quantenstrahlung. *Zeitschrift für Physik, 30(1)*, pp. 297–319.

Jordan, P. (1936). *Anschauliche Quantentheorie.* Berlin: Springer.

Jordan, P. (1971). *Begegnungen*. Hamburg: Gerhard Stalling Verlag.
Jordan, P. (1975). *Philosophie in Selbstdarstellugen I*. (L. J. Pongratz, Ed.) Hamburg: Felix Meiner Verlag.
Klein, M. J., & Toomer, G. J. (1979). Wolfgang Pauli, Wissenschaftlicher Briefwechsel mit Bohr, Einstein, Heisenberg u.a. (Sources in the History of Mathematics and Physical Sciences Ausg., Bde. 1 (1919–1929)). (A. Hermann, K. v. Meyenn, & V. F. Weisskopf, Hrsg.) New York, Heidelberg, Berlin: Springer-Verlag.
Kubli, F. (1970). Louis de Broglie und die Entdeckung der Materiewellen. *Archive for History of Exact Sciences, 7(1)*, pp. 26–68.
Moore, W. (1992). *Schrödinger Life and Thought*. Avon (GB): Bath Press.
Pauli, W. (1925a). Über den Einfluß der Geschwindigkeitsabhängigkeit der Elektronenmasse auf den Zeemaneffekt. *Zeitschrift für Physik, 31(1)*, pp. 373–385.
Pauli, W. (1925b). Über den Zusammenhang des Abschlusses der Elektronengruppen im Atom mit der Komplexstruktur der Spektren. *Zeitschrift für Physik*, pp. 765–783.
Pauli, W. (1957). Phänomen und physikalische Realität. *Dialectrica, 11*, pp. 36–48.
Pauli, W. (1984). *Physik und Erkenntnistheorie*. Wiesbaden: Springer Fachmedien.
Planck, M. (2001). *Vorträge, Reden, Erinnerungen*. Berlin Heidelberg: Springer Verlag.
Przibram, K. (1963). *Schrödinger, Planck, Einstein, Lorentz: Briefe zu Wellenmechanik*. Wien: Springer-Verlag.
Rechenberg, H. (2010). *Werner Heisenberg, Die Sprache der Atome*. Berlin, Heidelberg: Springer.
Rosenfeld, L. (1977). *Niels Bohr: Collected Works (Bd. 5)*. Amsterdam: New Holland.
Rößler, W. (2009). *Eine kleine Nachtphysik*. Reinbek bei Hamburg: Rowohlt Taschenbuch Verlag.
Schrödinger, E. (1926a). Quantisierung als Eigenwertproblem (Erste Mitteilung). *Annalen der Physik, 79*, pp. 361–376.

Schrödinger, E. (1926b). Quantisierung als Eigenwertproblem (Vierte Mitteilung). *Annalen der Physik, 81*, 109–139.

Schrödinger, E. (1926c). Quantisierung als Eigenwertproblem (Zweite Mitteilung). *Annalen der Physik, 79*, pp. 489–527.

Schrödinger, E. (1926d). Über das Verhältnis der Heisenberg-Born-Jordanschen Quantenmechanik zu der meinen. *Annalen der Physik, 79*, pp. 734–756.

Schroer, B. (2003). *Pascual Jordan, his contributions to quantum mechanics and his legacy in contemporary local quantum physics.* Retrieved from arXiv preprint hep-th/0303241

Schwartz, D. N. (2017). *The last man who knew everything: The Life and Times of Enrico Fermi, Father of the Nuclear Age.* New York: Basic Books.

Selleri, F. (1990). *Die Debatte um die Quantentheorie.* Braunschweig: Vieweg.

Stoner, E. C. (1924). The distribution of electrons among atomic levels. *The London, Edinburgh, and Dublin Philosophical Magazine and Journal of Science 48 (286)*, pp. 719–736.

Thirring, H. (1947). Erwin Schrödinger zum 60. Geburtstag. *Acta Physica Austriaca*, 105–109.

van Delft, D. (2014). Paul Ehrenfest's final years. *Physics Today, 67(1)*, pp. 41–47.

van der Waerden, B. L. (2013). *Sources of Quantum Mechanics.* Dover: Dover Publications.

von Weiszäcker, C.-F. (1984, November 23). Physik, knapp und klar, Paul A.M. Dirac 1902–1984. *Zeit.* Retrieved from https://www.zeit.de/1984/48/knapp-und-klar

Weinberg, S. (1992). *Dreams of a final theory.* New York: Pantheon.

GPSR Compliance
The European Union's (EU) General Product Safety Regulation (GPSR) is a set of rules that requires consumer products to be safe and our obligations to ensure this.

If you have any concerns about our products, you can contact us on

ProductSafety@springernature.com

In case Publisher is established outside the EU, the EU authorized representative is:

Springer Nature Customer Service Center GmbH
Europaplatz 3
69115 Heidelberg, Germany

www.ingramcontent.com/pod-product-compliance
Lightning Source LLC
LaVergne TN
LVHW010958250326
834688LV00003B/11